THE QUICKENING UNIVERSE

THE QUICKENING UNIVERSE

COSMIC EVOLUTION AND HUMAN DESTINY

Eugene F. Mallove

ST. MARTIN'S PRESS

NEW YORK

ILLUSTRATIONS BY THE AUTHOR

The artwork that illuminates this book was created by the author using an Apple Macintosh™ Plus computer running a program called *Full-Paint.*™ The author incorporated various public domain artwork using a *Thunder Scan*™ optical device to enter portions of existing drawings and photographs—modifying, combining, and reshaping them to suit his artistic fancy.

Design by Design Oasis

Library of Congress Cataloging-in-Publication Data

Mallove, Eugene F.
The quickening universe.
"A Thomas Dunne Book"
1. Cosmology. 2. Life—Origin. 3. Life on other planets. 4. Science—Philosophy. I. Title.
QB981.M256 1987 523.1 87-16150
ISBN 0-312-00062-6

First Edition

10 9 8 7 6 5 4 3 2 1

To my children,
Kimberlyn and Ethan.

To my wife, Joanne,
whose music sings of the stars.

To my parents,
who with love and devotion
brought me to the sea.

And to
Guy Murchie,
who showed the way.

One has been endowed with just enough intelligence to be able to see clearly how utterly inadequate that intelligence is when confronted with what exists. If such humility could be conveyed to everybody, the world of human activities would be more appealing.

—Albert Einstein, in a letter to Queen Elizabeth of Belgium, 1932

We cannot doubt the existence of an ultimate reality. It is the Universe forever masked. We are a part of an aspect of it, and the masks figured by us are the Universe observing and understanding itself from a human point of view. When we doubt the Universe we doubt ourselves. The Universe thinks, therefore it is.

—Cosmologist Edward Harrison, *Masks of the Universe*, 1985

A scientist lives with all reality. There is nothing better. To know reality is to accept it, and eventually to love it. A scientist is in a sense a learned child. There is something of the scientist in every child. Others must outgrow it. Scientists can stay that way all their lives.

—Biologist George Wald, remarks on accepting the Nobel Prize

CONTENTS

Preface and Acknowledgments

In the beginning, not a single word graced the glowing, almost frightening, blank screen of my word processor, but now here it is—the book that was yearning to be born. Appropriately, it is a personal reflection about life and meaning within a universe that emerged about 15 billion years ago from, perhaps, absolute nothingness.

Many streams of science and philosophy have flowed into this admittedly ambitious synthesis, and through its writing I have never forgotten the boundless debt I owe to scientists and thinkers, known and unknown, whose work fused in my mind. The writing consumed a few years of sporadic work, but the pace literally quickened in the final six months as completion grew deliciously reachable amid the frenzy of life's conflicting demands. To gauge the task more properly, almost twoscore years of study and questioning were necessary to sharpen these ideas. Since the book could not have been written a short twenty-five years ago—many key scientific ingredients not being known then—I wonder what the volume's sequel would reveal if I were fortunate enough to be able to set it down in 2010.

Many good friends were indispensable to my solitary labor on this project. A colleague in the writer's art, Ed Applewhite, provided a welcome boost by helping to bring my work to the attention of Thomas Dunne at St. Martin's Press. Literary agent Richard Curtis was always there to smooth troubled waters, as was Betsy Williams of St. Martin's Press. Nor can I forget Thomas Burroughs, who in 1983 as an editor at MIT's *Technology Review* magazine, struggled with the infant prose of a space engineer who was stubbornly bent on becoming a writer. How can I measure the depth of my gratitude to Joel Garreau and Dan Morgan of *The Washington Post* Sunday "Outlook" section? It was they who noticed my writing and thought it worthy enough—albeit with their editorial assistance—to publish at this writing a dozen major pieces in "Outlook," many of which have been adapted as parts of the chapters that follow.

I must credit my friend Robert L. Forward, who in the early 1970s spurred in me an already blossoming interest in the search for extraterrestrial civilizations and interstellar travel. That fascination was planted earlier by many wonderful authors, among the most memorable being Walter Sullivan *(We Are Not Alone)* and Carl Sagan and the late Joseph Shklovskii *(Intelligent Life in the Universe)*. Scientific colleague Gregory Matloff has been a constant and generous compatriot with whom far-ranging discussions have always inspired new ideas. My friend Steven Rubin, with his deep interest in the philosophy of science, has immeasurably increased an engineer's understanding of the mysteries of quantum mechanics. Over the past five years Steve and I have had innumerable noontime discussions at the York Steakhouse "School of Physics." And Professor Paul Horowitz of Harvard University contributed unwittingly by allowing me to join a thoroughly inspiring adventure as he occasionally relied on my elementary mechanical assistance to Project Sentinel—a major effort in the search for extraterrestrial intelligence that I describe in Chapter 11.

I first enjoyed the writing of Guy Murchie in his incomparable *Music of the Spheres*, a remarkably prescient gift from my dear Aunt Toby back in 1961. In my mind, this masterpiece was surpassed in 1978 only by his *The Seven Mysteries of Life*. I never tire of reading and rereading Guy Murchie's poetic adulation of life and the cosmos. We met first in 1985, discussed our "harmonious philosophies" (as Guy would say), and since have joined in many engaging conversations. *The Quickening Universe* would likely never have emerged without the inspiration of this delightful octogenarian who continues to chant the song of the universe whenever he speaks or sets pen to paper.

Last but certainly not the least of major influences have been anthropologist and international radio broadcaster Robert Arnold and astronomer Gerald S. Hawkins (author of *Stonehenge Decoded* and *Mindsteps to the Cosmos*). From our unique vantage point on this electromagnetically throbbing planet as former co-workers at the Voice of America and the United States Information Agency, we three have been privileged witnesses to the great mysteries of blending and clashing human cultures. Our wide-ranging conversations have helped enormously to expand my perspective on the course of human civilization.

I am also thankful to those intrepid souls who waded into portions or all of my first draft to hunt for errors: Gerald S. Hawkins, Robert Arnold, Guy Murchie, Sten Odenwald of the U.S. Naval Research Laboratory, cosmologist Marc Davis of the University of California at Berkeley, Professor Lawrence Fagg of the Department of Physics of Catholic University, exobiologist Lynn Griffiths of NASA, Steven Rubin of The Analytic Sciences Corporation, Richard Barnes of MIT Lincoln Laboratory, and philosopher Renée Weber of Rutgers University. Of course, I claim final responsibility for any errors of fact or attribution that may still remain hidden.

Finally, *The Quickening Universe* could not have been composed without the patience and understanding of my musically talented wife, Joanne. Perhaps because her own cosmic music is in such jeopardy in the present world, she could sympathize with the always risky business of trying to translate the music of science for a public that is not always so patient and receptive. But, dear reader, if you have managed thus far, I'm confident that *you* will enjoy a not too demanding voyage beginning with cosmology, passing through life, and on to the most remote destiny that we can imagine.

E. M.

Milky Way Farm
Bow, New Hampshire
1986

PROLOGUE

You must have chaos in your heart to give birth to a dancing star.
—FRIEDRICH NIETZSCHE

There was a star danced, and under that I was born.
—WILLIAM SHAKESPEARE
Much Ado About Nothing, II, 1.

Cosmic evolution, the endless transformation of all things through all time, is the common heritage and destiny of every subatomic particle and living being in the universe. On the blue planet Earth, third rocky orb out from the breathing Sun, the story is coming together, rippling through an emerging but still weak global consciousness.

The epic is not yet commonly known, nor is there agreement on its significance even by its many co-authors on this little world—each interprets his or her own meaning from the drama. The tale of cosmic transitions from the earliest times onward to the vast *to be* is even less familiar to the wider population. But if there is to be a sunny future for this local blossoming of cosmic awareness—mind out of matter—each new generation must hear the song of the universe that took centuries of investigation to compose. It is an unfinished melody and may always be so. Looking across the universe with our mind's eye, we end a disembodied hovering in the depths of space, journey past Sol, the dancing star that gave birth to

Human figure adapted from *Gulliver's Travels*, illustration by T. Morten.

life, and listen to the murmurs of a small being on a fragile but potent world. . . .

On the beach glistening white sand touches the sea, at the edge of the cosmic ocean in body and spirit. A boy in his teens contemplates the universe, reading of stars and space by noted astronomers. With symphonies of Beethoven echoing in his mind, he pauses to peer at the white surf crashing in from the cold Atlantic Ocean onto a beach in New England. Sand slips through his fingers, and he wonders with awe that all the grains of sand on the world's beaches do not outnumber the stars in the visible universe. Yet in a clump of sand no bigger than the end of his finger, there are more atoms than all those stars. These are passionate thoughts for a young mind beginning its reach to understand the cosmic order. How does this affinity for the depths of space and the innards of matter develop in a person, later to expand in harmony with the yearnings of others? Does this quest come from primordial stirrings that were long ago encoded in the genes?

There is a myth retold most dramatically in Arthur C. Clarke's movie *2001: A Space Odyssey* and other stories, that humanity was infected with cosmic fire—a seeking after distant worlds—by an older, wiser, and even magical race of beings that innumerable millennia ago engendered this spirit. A more stirring and perhaps more believable viewpoint is that life—generic "life" or organization, not the exclusive possession of this world—represents the only way for an otherwise meaningless universe to know itself. To that end, the quick were first parted from the dead. The universe came alive. Life, however originated, was invested with a message to seek new niches on the planet, to explore the order of the world, and later to seek distant environments in space. Life, beginning in Darwin's imaginary "warm little pond" or emanating from a crystal scaffolding in some craggy stream on Earth or near distant stars, carries with it an inevitable and eternal message to explore, to colonize, to know, and to transform itself.

Many people content with their earthbound business are unmoved by cosmic geography and don't often reflect on it. That they live on a dust-mote planet that orbits a very ordinary star in an equally ordinary galaxy of suns hardly affects their psyches. They conduct daily affairs within Earth's eggshell-thin atmosphere, unaware of the larger cosmic reality. But cosmic awareness is possible—cosmic consciousness, if you will—the ability not to lose for a moment a sense of smallness in the expanse of space and infinity of time. The feelings engendered by cosmic awareness may be similar to what some would call religious experience. They are mystical and transcendent feelings, but they stem from contemplating a marvelous natural order.

It would be difficult to surpass the expression of universal awareness revealed in the writings of Albert Einstein. In "The World as I See It" in 1934 he wrote of the cosmos: "A knowledge of the existence of something we cannot penetrate, our perceptions of the profoundest reason and the most radiant beauty, which only in their most primitive forms are accessible to our minds—it is this knowledge and this emotion that constitute true religiosity; in this sense and in this alone, I am a deeply religious man." Or from his "The Religious Spirit of Science" of that same year, ". . . [the scientist's] religious feeling takes the form of rapturous amazement at the harmony of natural law, which reveals an intelligence of such superiority that, compared with it, all the systematic thinking and acting of human beings is an utterly insignificant reflection." Einstein also said that "the most beautiful experience we can have is the mysterious" and that "the contemplation of this world beckons like a liberation."

The image of searching man held by Einstein and by Isaac Newton two hundred years before him is of a being wading in the shallows of the ocean of physical reality—trying to fathom the entirety by sampling only a part. We can go further into those waters and overcome our hesitation and at great risk dangerously suggest "purposes" of creation. Dangerous, because teleological thinking—attributing long-range "purpose" in nature—is fraught with philosophical traps and contradictions. But it isn't easy to ignore the evidence that we live in a quickening universe in which the "laws of nature," when traced through the corridors of time, are seen to spawn organized and evolving complexity. Subnuclear particles "inevitably" form atoms, atoms "inevitably" group into molecules, and molecules—we must believe—given a chance, quicken. Clearly, life is an emergent property of matter. Matter comes to life.

It can only be a guess—an article of faith supported by the latest cosmological findings—but the mysterious universe seems in some unfathomable sense "destined" to support organization and life, with the ability to reflect on itself and its relation to the unthinking part of the cosmos. How much of that "purpose" is programmed in the very structure of elemental matter? How much of that "purpose" later transformed its program to the more complex structures of primitive life and then again to the unimaginably more complex structures of the brain? Is the ultimate "goal" to ignite that cosmic fire that then bodily and mentally seeks the universe? These are mystical feelings, I am aware, but they seem to converge at a lightning-quick pace with the conclusions of modern science.

A tiny sample of molecular DNA, only twenty-six thousandths of a gram—a bare smudge on a spoon—has sufficient genetic extent to encode the heritable characteristics of the more than five billion human beings living today. Can it be true that this small DNA sample not only specifies the physical character but the scientific and artistic passions as well of all human beings? Is the message of the urge of physicists, biologists, starflight seekers, and space dreamers locked within the DNA? In a way yes, and in a way no. People have devised libraries, museums of art, films, plays, every manner of preserved and transmissible knowledge and sensation, all of which reside entirely outside the genetic code that each of us carries. This is a cultural transmission that is extragenetic—a cultural evolution infinitely more flexible and prone to remarkable invention. But it was, indeed, the very DNA—child of the physical laws—that ignited these libraries and art forms that transmit culture through space and time. This is how the otherwise barren universe comes to know itself here and similarly, perhaps, elsewhere.

What stage is next on the continuum of evolution from atom through library? Perhaps it will be the linking up of a universal awareness formed by the communicating network of species, independently evolved and culturally formed in diverse parts of the cosmos. Or will it be that magical state of the fetal starchild heralded in the movie *2001*, an ephemeral extradimensional existence? As cosmologists explore the fate of the universe, we come closer to discerning the ultimate destiny of life. Some speculate grandly that life could play a role that will affect the universe on a cosmic scale.

Whether terrestrial life meets with success in its striving or is consigned to an all too easy cosmic oblivion, we have reached an evolutionary stage at which the heart of a deep mystery is exposed but still locked. Why is the world so arranged that we can understand it at all beyond the bare needs of animal survival? How and why do we understand mathematics? And the same question pertains to the physical laws of the universe that are dimly perceived. As Albert Einstein said, "The most incomprehensible thing about the universe is that it is comprehensible." And from where does Man's affinity for music come—an apparently extravagant relic of evolution that is intimately linked with mathematics and physics? If we persevere long enough, answers may come to these questions.

Human beings individually have only a brief time in this world to form an image of the cosmos. Their minds are like film in a camera of awareness. Birth and death are the opening and closing of the shutter. Yet generations of striving to understand have led to a picture of the universe far more complete than any of us alone could have hoped to develop. We live in a germinating age of scientific exploration in which the exposure seems not hopelessly removed from highest possible clarity, though even that would fall far short of a vision of the perhaps infinite order beyond. I savor with delight that picture from which the tapestry of the quickening universe was woven.

LARGE AND SMALL NUMBERS WITHOUT FEAR

(Powers of Ten Made Simple)

Scientists use an easy-to-understand shorthand notation when speaking of very large or very small numbers. Such quantities appear often in cosmology and physics, so to save writing out large numbers of zeros that are hard to grasp quickly, I occasionally use this scientific notation. The rules are simple:

For numbers *larger* than 1.0:

The superscript tells how many zeros to put after the "1." Examples:

10^2 means 100
10^3 means 1000
10^4 means 10,000
10^{11} means 100,000,000,000
etc.

For numbers *smaller* than 1.0:

The superscript with a *minus* sign tells how many zeros are in the bottom (denominator) of the fraction. Examples:

10^{-1} means 1/10 or 0.1
10^{-2} means 1/100 or 0.01
10^{-3} means 1/1000 etc.
10^{-11} means 1/100,000,000,000

Not only is this notation compact, but one can see instantly which of two numbers is larger and by how many powers of ten. That's all there is to large and small numbers. Enjoy.

COSMOLOGY

Adapted from a photograph of the Very Large Array, Socorro, New Mexico.

1
The Silence and the Search

Our chief goal is a kind of self-knowledge as deep as our oldest myth: how it came about on this earth that the quick were first parted from the dead.

—Philip Morrison, 1985

My own view is that although we do not yet know the fundamental physical laws, when and if we find them the possibility of life in a universe governed by those laws will be written into them. The existence of life in the universe is not a selective principle acting upon the laws of nature; rather it is a consequence of them.

—Heinz Pagels, *Perfect Symmetry*, 1985

Knowledge of the cosmos is much more than a luxury for cultivated souls. It is the foundation of a cosmic consciousness. It casts light on the heavy responsibilities that have fallen upon us.

—Hubert Reeves, *Atoms of Silence*, 1984

We are children of the universe truly lost in space and time. A human life is as a blink of the eye to the time since everything began, and the estimated 15 billion years since then may turn out to be an infinitesimally smaller moment against what remains. On the scale of the cosmos the *mass* of a person is geometrically midway between that of an atom and a large planet's heft. And as for comparative *dimensions*, the size of a planet is the geometric midpoint between an atom's breadth and the expanse of the visible universe. That is, an atom's size bears about the same ratio to a planet's diameter that the latter bears to the most distant objects that are possible to observe.

The scale of a human body is far less to a planet and even less to the cosmos than one of the intestine's essential microbes is to the whole body. Human dimensions are literally microbial only when compared to something as small as a mountain range. We are strangely much nearer to atomic dimensions than to cosmic distances. The corporeal self is scarcely more than an overgrown dynamic "crystal" to the atoms within. Dimensionally compared to the universe, we barely exist.

Long ago the universe was invisible and silent, for there were no eyes to see, no ears to listen to its growing rhythms: a universe with no senses or self-apprehension, only dead matter drifting, time and space expanding. We believe this but find ourselves struck with paradox: a universe that did not know itself then, but that later awoke in a springtime of passion to know its history and the future. Did this really happen? Modern science, always mindful of the possibility of error, offers an emphatic "yes."

To a civilization in its infancy, the cosmos is a profound loneliness more chilling than the isolation of astronauts newly arrived on the barren Moon. We humans are the only examples of all that we are—at least as far as we have any proof. And therein lies the source of our millennial conceit and tides of woe. Gradually we are discovering greater intelligence, consiousness perhaps, in and among the so-called lower lifeforms—our distant relatives and fellow planetary travelers. We stand amazed and conceitedly imagine that *we* have become the undisputed pinnacle of sentient life on a planet of clacking, hissing, and babbling incomprehension. The vast space between stars and galaxies seems to insulate us from suspected kin on far-removed worlds. In moments of self-delusion we have the luxury—or the misfortune—to believe that we are alone with our passions and thoughts in a barren cosmos.

So a long time ago we invented deities—first lesser gods, inhabiting the mountains and oceans, with whom to discourse and do battle. Then our imagination stirred and conjured transcendent God, "living" beyond space and time but intervening miraculously in human affairs from time to time. A fostering spirit, a father or mother figure, helped and still aids many to comprehend the loneliness of this world.

The explosion of science in the last five hundred years has converted much of that religious spirit to skepticism. We have become mostly Copernicans and Darwinians, whether secret or confessed, and believe that we have no special place or status in the universe, no anointed spot in cosmic geography, and no prominence in a universe likely to be teeming with older and wiser races. To be sure, there is room to recant—to mistake the absence of evidence for evidence of absence in the matter of intelligent extraterrestrial life. Some even in the halls of science harbor the now heretical view that life must be a unique or extremely rare phenomenon in the universe. And so the silence of our cosmic isolation continues.

Within a thousand years for sure, and perhaps in much less time—perhaps even within this century—it will be all over. We will probably have discovered that the universe gave birth to life and intelligence in more than one locale—perhaps in vastly more than one. We cannot prove this now, but as science grows in understanding of matter, chemistry, and life itself, the more apparent it seems that "life" is the inevitable offspring of senseless matter—given the proper environment. The universe quickens and looks back on itself with a smile. It has come alive. Magically, quickening means not only literally "coming to life," but also describes the accelerating pace of evolution. Quickening is also the stage of pregnancy in which the movement of the fetus first can be felt. So it seems highly appropriate to suggest that the laws of physics are literally *pregnant* with life—complexity. They set the stage for life, though not necessarily a specific kind of life.

Cosmology "ordains" life? The laws of physics "ordain" life? How rash a thought, much too reminiscent of a divine pronouncement! Yet science is certain that the laws of physics ordain atoms and atoms in turn molecules, stars, and planets. There are veritable libraries of scientific supporting arguments for the claim that the constitution of matter ordains life, but precious few who will draw the conclusion so sharply.

"Yes, of course there are probably millions of extraterrestrial civilizations or at least many worlds on which primitive life has arisen. But the laws of physics ordaining life? That's too extreme—too mystical." Something is holding us back from the logical inference that the laws of physics promote life as well as atoms, perhaps because of our temporary difficulty in fathoming the chemical evolution that led to life. As an example, no one would call the simple chemical combination of hydrogen and oxygen to produce water a mystical phenomenon, yet there is a certain "inevitability" to the reaction. Put the two elements together and one gets water of combustion. It is only when we go on to speak of "inevitable" chemical reactions that lead to *life* that scientific hackles seem to be raised. Or is the problem that until recently the laws of physics had no real history? Now increasingly, physical theories are intimately linked with the details of the evolving early universe. The laws have a past and a future; they aren't static in the present. A grand continuum of cosmic evolution leading to and beyond life seems a natural outgrowth.

The Copernican principle is a two-edged sword. On one hand terrestrial life has no special location in the universe. Earth is not the center of the solar system; our whirl of planets is a tiny outpost in the Milky Way galaxy with at least 100 billion stars and perhaps as many as 400 billion; and the Galaxy is only one of hundreds of billions of other galaxies in the visible universe. But this has been mistaken too often for the idea that life and intelligence are not very special manifestations of matter. We should be careful to remember that size, position, and duration are not the only measure of merit. No matter how small or seemingly weak, there is unquestionably a special grace to a fragment of the universe that aspires to comprehend the whole. We can view "intelligence" as the enfolding of physical laws to form organizations of matter and energy capable of discerning those same governing principles that gave birth to intelligence. In the deepest sense we are examples of the physical laws doubling back on themselves for self-examination.

Enlarging this view, we suspect that the universe may harbor more guises of life than are dreamed of in our natural philosophy. Science fiction may be a more proper guide than science to the multitude of ways in which the universe comes to know itself. We are now mere provincial terrestrial biologists with one example to study of what we call life. Could life be a more general principle in the universe that encompasses many forms of organization striving to sustain and multiply itself, reaching ever more complex and advanced states of awareness? No one yet can prove that life is a more general drive of matter and energy than mere creeping carbon-based creatures struggling to emigrate from a sticky planetary surface. Science fiction writers have already given us subatomic beings living on the fertile nuclear surfaces of so-called neutron stars, an electromagnetic intelligence that spreads across millions of kilometers in deep space, and other life-forms too varied to recall. What other-wordly life have the wizards of time and space not foreseen?

The entire history of science is like a blinding flash of human awareness of the universe. The scientific enterprise is how life goes about building testable models of reality. In fact, those testable models define "reality," because the traditional guide of "common sense"—naïve realism—can no longer substitute for theories of relativity and quantum mechanics to explain important phenomena. At any stage in its development science may be incomplete and even wrong in its descriptions, but by common agreement its theories are always subject to revision and expansion.

Science is the antithesis of dogmatic belief, yet is not without its peculiar dimension of faith: the credo that the underpinnings of the universe are capable of rational interpretation. Science is neutral to preconceived philosophical ideas, but it is not immune to them. The direction of science in this bit of the universe is charted by passionate beings with a taste for beauty—each with an emotional sense of how the world should be put together. They make choices every day that guide the searchlight of understanding over the dark territory of ignorance. There is ample opportunity for error, but also for self-correction.

The search for life's place in the cosmic scheme proceeds along diverse fronts. In 1969, people of Earth first grasped the soil of the Moon and found it wanting in organic complexity—no life there. Then in 1976 two robot spacecraft, Vikings 1 and 2, landed on Mars to sift its substance for microbes, and to peek with cameras at its surface for the hoped-for but fleeting glimpse of an unexpected "Mars mouse" scurrying among the rocks. Alas, no Mars mouse or even a Mars microbe. At the two widely separated landing sites, and insofar as our technology knows how to recognize something with the general character "life," Mars proved to be a dead planet. Not all scientists, however, would agree with the negative conclusions of the Viking experiments. Some suggest that it is absurd to imagine that tests at two isolated landing sites represent a thorough search for life on Mars. Some believe that the sensitivity of the tests was not great enough to warrant pessimism. But the scientific consensus about this once promising and possibly oceanic world, is that life never arose or once did but has long since disappeared. The final judgment about Mars will come within the next twenty-five years when human expeditions are sure to be launched to prospect for life firsthand.

Earth and Mars do not exhaust the possible abodes of life even within the solar system, though relatively clement Mars seems to have had the best chance for life apart from Earth. Perhaps we will yet find some kind of life floating in the atmosphere of Jupiter, swimming in its moon Europa's ice-covered ocean depths, or struggling on the Saturnian moon, Titan—the only satellite in the solar system with a substantial atmosphere. Finding any kind of life independently evolved within the solar system would augur well for its prospects far from the Sun near similarly temperate stars. Not finding it would be a disappointment, but would not necessarily mean that life was a rare occurrence in the cosmos.

Since 1960, astronomers have searched with radio telescopes for signals from other civilizations in distant solar systems. It is not surprising that so far they have not discovered any artificial signals that cannot be attributed to the natural burps and beeps of the cosmos. The search, until recently, has been far too sporadic for scientists to have expected success, but with the advent of inexpensive and powerful new computers coupled to radio telescopes, the chances are improving. Yet there is still an enormous cosmic haystack to search to find the needle signals. There are billions of frequency channels to scan and countless directions in space toward which to point our electronic ears. Unless they are very lucky at the outset, decades will pass before astronomers tire of the hunt for strong, deliberately beamed signals. If none have been found, they have plans to launch larger antennas into space, where, shielded from terrestrial radio interference, they will try to tune in on weaker "leakage signals" from putative extraterrestrial domestic broadcasts.

In the laboratory biochemists tinker with the molecules of life and their constituents, hoping to trace the route first traveled by matter on the road to terrestrial life. Even though our power to control cellular machinery through genetic engineering grows stronger, science is still at the stage of alchemy in understanding how lifeless matter at least 3.8 billion years ago transformed itself to living things. Progress is being made and new ideas continue to infuse the search, but an enduring mystery shrouds those ancient stirrings that led to us. What spark ignited the fire of life in Darwin's "warm litle pond"? Or was life carried to Earth as interstellar seeds—accidentally arrived or deliberately sown by an ancient race among the stars? Even if that panspermatic vision were true, the mystery of that earlier chemical evolution that produced the benevolent aliens would remain.

The great nineteenth-century Russian visionary of space travel Konstantin Tsiolkovsky regarded Earth as the "cradle of life." But the true womb of life is the entire universe—the totality of existence that we do or can experience. Our life-giving star, the Sun, and its retinue of planets owe their presence to an evolution many times older than the chemical tournament that led to life as we know it. Life science and cosmology are inseparable studies and are becoming more so as time passes. The dependence of life on the evolution of galaxies, stars, and planets is absolute. But the evolution of space, time, and matter in the earliest phase of the expanding universe is even more paramount. All that will ever be called life depends most fundamentally on elementary particles and their forces, and even the basic dimensionality of space. Our understanding of the microcosm's rich hierarchies is improving each year and is leading to an all-inclusive framework of physical law. That wondrous structure is the true cradle of life.

The present age is most remarkable, and we should count ourselves incredibly fortunate to be living at this one special moment. Just as we have stepped off the planet and begun to test the ocean of space, just as we begin a promising radio search for extraterrestrial intelligence, and precisely as we begin to master the genetic machinery of terrestrial life—at that very time science has opened wide the door to a comprehensive physical theory of existence. Theories of the expanding universe are rapidly converging with unified theories of force and matter at the level of subatomic particles. The most intriguing part of cosmology—the first fraction of a second of cosmic history—has become indistinguishable from the once separate speciality, particle physics.

The barrier of the unknown that predates the "Big Bang" expansion of the universe has now been breached. Science is now knocking loudly on Heaven's door, seeking, as did Einstein, to "know what was in the mind of God" 15 billion years ago. Some physicists, awestruck by the numerous improbable physical coincidences that allow the universe to exist and to support life, have proposed theories to explain those fortuitous circumstances. Their work is forcing them to ask the ultimate philosophical question, long the province of religion: did the universe come into being by chance or by "design"? What in fact does chance or design mean for the cosmos, which by definition includes everything and for which only one example can exist?

Physicists are quite literally "playing God," delving into theories of "metalaw"—laws that may describe how the physical rules were or were not forced to be what they are. Particle theorist and Nobel laureate Steven Weinberg described the state of cosmological knowledge in 1985: "We can trace the history of the present period expansion back to its first million years, or its first three minutes, or its first ten-billionth of a second, but we still do not really know if time really began just a little before then, or if so, then what started the clock. It may be that we shall never know, just as we may never learn the ultimate laws of nature. But I wouldn't bet on it." The origin of the universe is no longer the most fundamental problem. In Weinberg's words, "We want to know the origin of the rules that governed the universe and everything in it."

But science has been fooled by its apparent power to explain almost everything before. At the close of the nineteenth century prominent physicists such as Lord Kelvin in England were convinced that physics—admittedly the most fundamental science—was rapidly becoming a closed subject. They thought almost everything worth knowing in physics had been discovered. Then the atom opened up, the shock of relativity and later quantum mechanics; a veritable universe of new ideas blossomed. Science was chastened by that bad prediction and will be more careful in its optimistic pronouncements about unified theories. But looking back, there seems not to have been a time when a broader scope of cosmic history was confidently patched together than right now. Connections between once diverse sciences are emerging and filling in blank areas on the painting of cosmic evolution. With some pretension, we might call the anticipated linking of cosmology, particle physics, chemical evolution, and biology a "Grand Evolutionary Theory."

Implicit in the search for extraterrestrial life is the belief that "the universe never does anything just once." From this perspective the universe would seem fantastically imbalanced to have "accidentally" originated life on but one very receding planet. Scarcely admitted by the advocates of a universe filled with life is the almost essential concomitant theory: the "plan" to make dead matter quicken is inherent in the laws of physics. It is almost a truism that if one believes in the prevalence of life in the universe, then one must also believe that the laws of physics *imply* and in a sense "mandate" life. Scientists working on the problem of extraterrestrial life more often state that life is a cosmic accident, but one that may happen frequently in suitable environments. The distinction is exceptionally fine, however, between prevalent accident and physical inevitability. Only an exceedingly stubborn philosopher would be willing to split that hair. If life emerges frequently, even if not inevitably, in suitable environments, then we should think of life essentially as an extension of the laws of physics.

As we learn more about the laws of the universe and begin to collect evidence of organization on other worlds, the view will likely grow that life was no accident, that there is a destiny in matter that will not be denied. Science will have no proof of this until the silence is broken—until an alarm in the night tells us we are surely not and have never been alone. The sound that rudely interrupts Earth's isolated complacency may soon be heard.

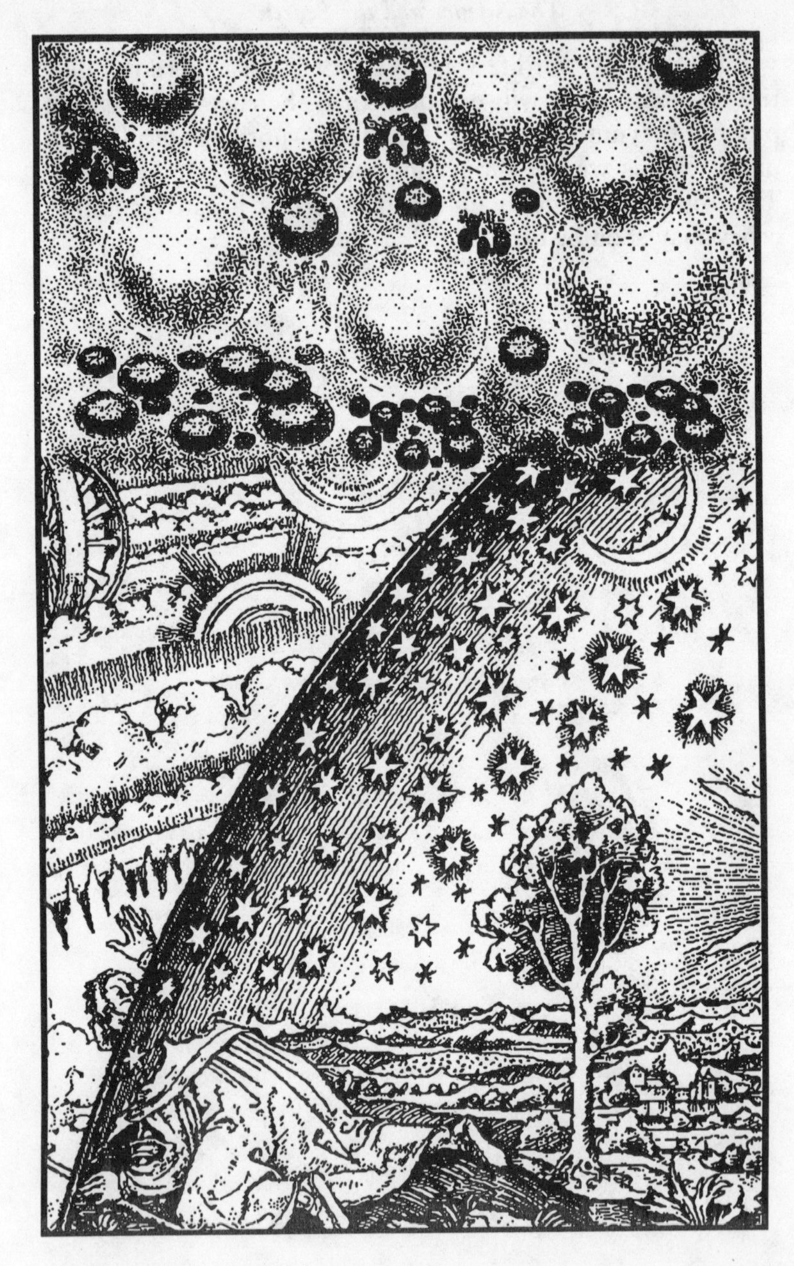

Adapted from a nineteenth century woodcut illustrating a medieval cosmology.

2

THE NOTHING UNIVERSE

The inflationary model of the universe provides a possible mechanism by which the observed universe could have evolved from an infinitesimal region. It is then tempting to go one step further and speculate that the entire universe evolved from literally nothing.

—ALAN GUTH AND PAUL STEINHARDT, 1984

The nothingness "before" the creation of the universe is the most complete void that we can imagine—no space, time, or matter existed. It is a world without place, without duration or eternity, without number—it is what mathematicians call "the empty set." Yet this unthinkable void converts itself into the plenum of existence—a necessary consequence of physical laws. Where are these laws written into that void? What "tells" the void that it is pregnant with a possible universe? It would seem that even the void is subject to law, a logic that exists prior to space and time.

—HEINZ PAGELS, *Perfect Symmetry*, 1985

Existence, the preposterous miracle of existence! To whom has the world of opening day never come as an unbelievable sight? And to whom have the stars overhead and the hand and voice nearby never appeared as unutterably wonderful, totally beyond understanding? I know of no great thinker of any land or era who does not regard existence as the mystery of all mysteries.

—JOHN ARCHIBALD WHEELER, 1986

We cosmic questers are born into this world and all too soon exit across the threshold of life. This is deeply puzzling: why should we exist at all? Before our conception and birth the world presumably endured without us. And certainly not very far in the cosmic past Earth traveled its celestial course bearing not a single hominid, neither hairy nor naked. So strange is the awareness of mortality that we might sympathize with, yet not accept, the extreme solipsist belief that when one dies, the world goes out like a light.

Now take the question of existence one step further. What "was" before there was a universe? The question would be meaningless if the universe had no beginning. Until quite recently—a little more than twenty years ago—it was still possible to theorize that the cosmos had always existed. Of course, the idea of an infinitely old universe has its own paradoxical flavor. But, as we shall see, it is no longer scientifically easy to cling to that concept. So what came "before" the universe? The answer, it seems, may be absolutely nothing. Now there are surprisingly many varieties of "nothing," and we must be careful to define the kind of nothing when we speak of nothing!

There was a time, perhaps 10 to 20 billion years ago, when all of the observable universe did not exist. At that "time," in fact, neither space nor time itself was in any form with which we are familiar. Perhaps there was literally nothing, not even a "when" or a "someplace." These concepts, admittedly hard to grasp, are only some of the mind-wrenching but nearly inescapable conclusions that have been drawn in the twentieth century by cosmologists—astrophysicists who study the large-scale structure, origin, and evolution of the universe.

New theories that unite the fundamental forces of nature, combined with observations made by the world's largest telescopes and nuclear particle accelerators, inexorably press the frontiers of physics toward answering questions once the exclusive province of religion: How did the universe begin—if in fact it did have a beginning? Why is the universe as it appears today? What is the ultimate fate of the cosmos? How the universe delivered life and intelligence from its womb depends intimately on cosmology.

The most dramatic recent course of science has been the

melding of the structure of the macrocosm—the realm of stars and galaxies—with the microcosm of subatomic particles and waves. Today astronomers make observations that directly bear on theories of the subatomic order. Conversely, nuclear particle accelerators reveal facts with immediate astronomical significance. Virtually all cosmologists believe that the universe sprang forth in what is known as the original Big Bang explosion from a state of extraordinary compression, phenomenally high temperature, and unification of fundamental forces. They believe that all the disparate contemporary forces of nature behaved as a single force for a fleeting instant. But many questions remain unanswered, such as why the universe of galaxies is as uniform as it appears, if its original state was chaotic. Or if the earliest state wasn't disorganized, but instead was perfectly uniform, how did manifestations of nonuniformity from galaxies to clusters of galaxies come about? How did matter predominate over antimatter, and why is the universe so precariously perched between being conducive to life and making life impossible?

Surprising progress has been made in answering some of these questions. Alan Guth, a young theoretical physicist now at M.I.T., in 1980 developed the first of what are called inflationary theories that describe how the original Big Bang explosion was energized. The latest thinking doesn't rule out, and may even make more plausible, the existence of other universe domains—realms that would be forever beyond our ability to investigate. These would be mere bubbles that started off in a froth of fluctuating primordial fields, much as our own tiny region of a local bubble of space and time, what we call the observable or visible universe.

Moreover, the largest part of our own cosmic bubble that we may hope to see—because light hasn't traveled far enough since the beginning of time from more remote quarters—may be only a minute portion of a much greater realm encompassing it. When we look at light coming from a great distance, we eep into the past of that region of space. We can look out ıs far as light has had time to travel since the universe began to expand.

The Expanding Universe

The enormous scale of the universe was not recognized until the 1920s, but prior to that era many had begun to imagine a fantastically larger cosmos beyond the "nearby" stars. William Herschel, the eighteenth-century British astronomer and discoverer of the planet Uranus, had speculated that the fuzzy patches of light in the sky seen through telescopes were "island universes"—much like the aggregation of stars, the Milky Way, that the Sun appeared to inhabit. The German philosopher Immanuel Kant of that period espoused the same view. But the truth of the idea was not confirmed until American astronomer Edwin Hubble with the 100-inch Mount Wilson telescope was able to resolve individual stars in the Andromeda nebula. We now realize that this impressive whirlpool of stars lies about two million light years away and has a spiral shape that would mimic our own galaxy if we could look back at the Sun's home from afar.

The Milky Way galaxy with its up to 400 billion stars is now seen to be but one of hundreds of billions of other galaxies that abundantly populate astronomical photographs. It is a sobering thought that light, speeding at approximately 300,000 kilometers per second, would take almost 70,000 years just to traverse the diameter of our pancake-shaped Milky Way. Yet even nearby galaxies like Andromeda are millions of light years away. Light now splashing on Earth from Andromeda began its journey when prehuman hominids trod this world. Not terribly far from the edge of the Milky Way galaxy, embedded in one of its spiral arms, the Sun circles the galactic center once each 200 million years. At the Sun's last birthday, figured in these galactic "years," dinosaurs were celebrating in swamps.

By measuring the shift in the expected light spectra of distant galaxies, Hubble found that these huge islands of stars were receding from us with speeds of hundreds and even thousands of kilometers per second. The stars that comprise a luminous galaxy emit light over a range of frequencies, another word for different colors. By separating those frequencies in galactic light into a rainbow of colors with a device analogous to a glass prism, astronomers see patterns of lines caused by atomic emission and absorption of light. These are characteristic of particular elements like hydrogen or sodium when they are hot. If the line patterns appear to be shifted higher or lower from their normal positions on the frequency spectrum, this indicates that the galaxy is moving rapidly away from or toward the observer, depending on the direction of the lines' shift. This useful phenomenon, called the Doppler shift, is illustrated by the rising and then falling pitch one hears from a speeding train, whistling as it approaches and then recedes.

The spectrum of Andromeda, because it is shifted slightly toward the blue end, reveals that the galaxy is actually drifting 90 kilometers per second toward our Milky Way. Local deviations do occur from the overall pattern of galaxies receding from us and having their light shifted generally toward the red end of the spectrum. Those nearby galaxies dance about one another, drawn together by gravitational attraction. But overall, the universe appeared to Hubble to be expanding, not preferentially away from the Milky Way, but like bread expanding in an oven—each galaxy playing the part of a raisin in the loaf. No matter which raisin an observer inhabited, the expansion would look the same (ignoring the question of what raisins near the surface of the bread would see because the universe has no "surface").

The implications of an expanding universe were staggering.

Expansion meant that at one time all matter must have been compressed together with enormous density. To recognize this required a simple extrapolation backward in time. Evidence for expansion suggested that the comforting idea of an eternally unchanging universe, perhaps infinitely old, would have to be replaced by an evolving cosmos of finite age.

Resistance to the idea of a Big Bang explosion that set the universe into being was building, however, due to observational errors that implied a contradiction: that the universe was only two billion years old—younger than some rocks whose age was reckoned by measuring elements that had radioactively decayed. In the late 1940s Herman Bondi, Thomas Gold, and Fred Hoyle proposed a "steady state" cosmology that postulated that the universe had always existed and would continue expanding forever, while always looking the same (on average) to any observer. For the theory to work, they had to invent the then bizarre concept of spontaneous creation of matter out of nothing in the vast voids between galaxies. Only a few atoms per many cubic light years of space would have to appear each century for their model to have worked. The steady state theory presupposed neither a beginning nor an ending of time. The universe always had been and forever would be.

The steady state model was elegant and, in a way, very philosophically pleasing. The abrupt creation of the cosmos in an instant of time was avoided, and life was free to evolve on planets through eternity. The Sun might run out of fuel in some billions of years, but there would always be new stars to take its place—forever. But the virtual death knell of the steady state cosmology came with a serendipitous discovery by two scientists at Bell Laboratories in New Jersey in 1965.

Arno Penzias and Robert Wilson were reconditioning a large microwave horn antenna formerly used for communication satellite experiments, but which was now set up to measure the background radio noise coming from speeding charged particles in our galaxy. There was a persistent low-level electronic hiss which they could not remove, no matter what they did or in which direction they pointed the antenna. For a while they even suspected that pigeon droppings in the antenna had something to do with their problem. They and others soon realized that the omnipresent noise was the remnant birth cry of the Big Bang explosion echoing through space and time. They had detected a uniform background radiation of microwaves (like those in a microwave oven or from a radar) that had cooled (lowered in frequency) to a temperature of only 3 degrees Kelvin above the temperature scale's absolute zero as the universe expanded from a time when it had been much hotter. (A degree Kelvin has the same magnitude as a Celsius (C) degree, but the Kelvin scale begins at absolute zero, about 273°C below the freezing temperature of water, which is 0°C.)

The temperature and uniformity of the radiation fit perfectly with the model of a Big Bang explosion that had occurred 10 to 20 billion years earlier. The discovery, which later won Penzias and Wilson the Nobel Prize for physics, was too difficult for steady state cosmologists to integrate successfully into their theory. The almost indisputable proof that the universe did have a beginning that started with a staggering compression of everything must certainly be regarded as one of the most spectacular realizations of all time.

If it does nothing else to our psyches, the idea of the Big Bang may engender a peculiar claustrophobic feeling with the awareness that all of this spacious existence—oceans, trees, planets, people, stars—was at one time shrunken down and confining. Nor is that just a feeling. The early universe *was* sweltering and restrictive, because conditions then were so inclement, hot, and pressured that no organized matter as we know it today could have existed. It is extraordinary that we can retrace time step by step and arrive at an era when everything must have been squeezed on top of everything else. To understand the nature of the universe in that barely imaginable state is one of the greatest challenges science has ever faced.

One incorrect popular notion about the expanding universe is that matter somehow exploded from a very small locale *into* a larger or infinite space when the Big Bang started. The proper visualization is instead that all of space and the matter and radiation it contained expanded from a small point—perhaps *literally* a point, which is to say nothing at all. Space itself did not exist; it had to blossom. Cosmologists speak of the point as a "singularity," meaning a strange point in the fabric of what is called four-dimensional space-time. It represents the only thing that can be called an edge of the universe, literally the termination of time and space.

Bear in mind that it is not possible to relate this kind of expansion to anything in our daily experience as three-dimensional creatures who are hurtling forward in time—often called the fourth dimension. The best though imperfect image might be that of a bubble of space expanding outward from a minute point. Think of this bubble of space in some sense mixed or blended with time—that mysterious fourth dimension. "Outside" the bubble one should have the admittedly impossible mental picture of an infinite sea of nothingness—a sea which our limited minds endlessly seek to turn into "infinite space," an incorrect image.

The bubble universe might be finite, like the world inhabited by flat creatures living on the surface of an expanding balloon. All of their two-dimensional space could expand from a time of very great curvature (a tiny balloon) to a time of much smaller curvature (a highly inflated balloon) and the creatures would be hard-pressed to imagine their two-dimensional world "expanding *into* something." If Einstein's so-called space-time—the admixture of time and space—is finite, with a curvature such that a hypothetical space traveler would circle back to her starting point no matter how she tried to hold to a "straight line," then we are three-dimensional creatures who have the same conceptual problem as the flatlanders. Cosmologists refer to this as a "closed" universe.

But most theorists believe that the cosmos may be infinite. They speak of an "open" universe in which traveling on a straight line does *not* bring the explorer back to the starting point. Whether or not the universe is infinite, we must imagine the Big Bang as "an explosion everywhere" in infinite or finite space-time. In an almost unfathomable sense, every cubic centimeter of the universe was the site of the Big Bang. "Surrounding" the probably infinite domain of the universe, we should not even visualize "nothingness." Perhaps there may be other infinitely large universe "bubbles" lurking "outside" our own—a mind-boggling foam of universes, each with its own scheme of being which most likely will remain forever unknown to intelligence in our universe. A true infinity of infinities.

Theoretical physicist Steven Weinberg provided one of the most apt descriptions of the Big Bang in his book *The First Three Minutes:* "In the beginning there was an explosion. Not an explosion like those familiar on Earth, starting from a definite center and spreading out to engulf more and more of the circumambient air, but an explosion which occurred simultaneously everywhere, filling all space from the beginning, with every particle of matter rushing apart from every other particle." Weinberg's description of the Big Bang holds for a finite or an infinite universe. Galaxies don't just happen to be fleeing from one another; it is the expansion and stretching of the rubbery, fluidlike space-time that drags them apart.

When radio astronomers look at the 3°K cosmic microwave radiation with their dish-shaped antennas, they are really measuring the temperature of a very early phase of the universe—but a temperature *reduced* in proportion to the expansion since that time. The radiation comes from distant space and therefore a remote time in the past when the cosmos was only a few hundred thousand years old. The radio astronomers are observing a part of the universe as it was back then. In turn, the part of the universe that we inhabit sent out intense radiation long ago from its fiery youthful body toward equally distant parts of the universe which may have beings looking at *our* microwave radiation. The expansion of the universe has caused a "reddening" or lowering of frequency of the primordial light.

When the universe was for example 1,000 times more compressed, its temperature was 1,000 times hotter or 3,000°K. That was the epoch from which the present microwave background radiation derives—the era during which the first complete atoms formed and in which the radiation was last scattered. Applying only the most meager assumptions about the interactions of particles and radiation, and projecting earlier temperatures in a similar way, cosmologists have derived a model of the newborn universe that is confirmed by many astronomical observations.

One of the most noteworthy features of the visible universe today is the predominance of hydrogen and helium in the composition of matter. Helium makes up about 25 percent of the visible (light-emitting) massive particles in the universe (reckoned by mass, not by the number of atoms). Hydrogen constitutes all the rest. There are only trace amounts of heavier elements like carbon, nitrogen, oxygen, and metals, almost all of which were synthesized in cataclysmic explosions of massive stars called supernovae and in the exhalations of more moderate stars over the long course of their lives. If these stars had not sacrificed themselves to deliver the heavier elements to the interstellar stew, there would be neither rocky planets nor beings on them to speculate about all that came before.

The present element abundances cannot be explained by synthesis of helium from hydrogen in the thermonuclear furnaces of stars—a far too lengthy process for the amounts of helium observed. But the billion-degree temperature of the universe when it was only three minutes old would have "cooked" just that helium abundance from the original hydrogen nuclei—simple protons. This dramatically confirms a prediction of the so-called standard model theory of the Big Bang that is beautifully consistent with the microwave background measurements.

As reconstructed by theoretical physicists, the universe began some 10 to 20 billion years ago in a fiery sequence of events that transpired in only a few minutes. Some of the most fundamental cosmic history occurred within the first second! What was the universe like in the first moments of the Big Bang? The earliest time for which the known laws of physics have any meaning is 10^{-43} seconds after the ultimate beginning or imaginary "zero of time." According to present knowledge, this so-called Planck moment might be considered the beginning of the universe instead of the true "zero time." Inside the Planck moment lies the singularity of space and time, the nature of which is literally a hot point of contention among cosmologists. It may be analogous to the singularity that must inhabit the center of a black hole—a region of space-time that has collapsed, the result of too much matter aggregating in one spot. Unfortunately for present understanding of the primeval singularity, physicists have yet to develop a completely satisfactory theory that links Einstein's theory of gravity, space, and time—general relativity—with quantum mechanics, the theory that describes probabilistic behavior in the microworld.

The New Inflationary Universe

In 1980 Alan Guth conceived a breathtaking new view of how the Big Bang was set in motion shortly after the Planck moment—an idea that has become known as the "inflationary model." Instead of assuming that the expansion rate was relatively steady all the way back to the Planck moment, he postulated an exponentially expanding universe starting at 10^{-35} seconds. The mind-numbing rapidity of this exponential expansion, which occurred virtually instantaneously, consuming only 10^{-30} seconds, powered the universe into being and was then over. In a minute fraction of a second, the universe expanded from far below subatomic dimension by a huge factor—perhaps 10^{50} times or more. Inflation promises to show that the present form of the universe is the consequence of basic physical laws operating in the first fraction of a second.

At this point the delirious cosmic voyager may be justifiably puzzled by the word "size" applied to a possibly infinite universe. There are really two sizes of which cosmologists speak. One is the "observable" or "visible" radius and is defined for *any point* as the farthest distance from which light has had time to travel *to that point* since the beginning of time. Today the visible radius is therefore about 15 billion light years—light from farther away hasn't yet had time to reach us or anyone, anywhere. Far back in cosmic time, just at the end of the inflation period, the visible radius was only 10 centimeters! So the visible radius is constantly growing as the universe expands. With telescopes powerful enough, more and more of the universe becomes accessible, like the growing awareness that a child has of the mysterious world of adults.

But the other measure of the universe's size, called the *horizon distance*, is spectacularly larger than the mere visible radius—particularly for the inflationary universe in comparison to the steady growth or "standard model" of the Big Bang. The horizon distance, which also continues to grow, measures the farthest parts of the universe that can have physically influenced one another at any time. In the standard model of the Big Bang the visible and horizon distances have long since grown to be the same. But in the model of the universe boosted into existence after inflation, the horizon distance is a staggering 10^{25} or more times larger than the visible universe! The startling conclusion: all that we see or can ever hope to see is as nothing compared to what lies beyond. The visible universe is proportionally like an atom compared to an object at least 10 times larger than the solar system. The inflationary universe, which the majority of cosmologists embrace, is the proverbial atom in some giant's cup of tea.

Guth's inflationary early universe and more refined *new* inflationary theories that others (notably Paul Steinhardt, Andreas Albrecht, and Andrei Linde) have developed explain with one breathtaking vision at least three major problems that had been plaguing cosomologists. The first is the remarkable homogeneity of the cosmos on a large scale and the observation that it is isotropic—it appears the same to an observer looking in any direction. On a large enough scale the density of the universe as reflected in the cosmic background radiation is uniform to one part in 10,000 in all directions. One would have expected zones of very disparate density coming from an initially chaotic "soup" of particles. How could widely separated regions of space look so similar if they were too far apart—their mutual horizon distance too large—to have influenced one another since the beginning of time? What blending mechanism generated large-scale uniformity? Yet the distribution of matter in the universe is not nearly as uniform as the background radiation. How did galaxies and clusters of galaxies emerge from primordial homegeneity?

The second mystery is the lack of evidence for monstrously large, complex particles, called magnetic monopoles, that according to prevailing theories of force unification should have been as numerous as hydrogen atoms. A magnetic monopole would have the mass of a bacterium—colossal by elementary particle standards. The third and perhaps greatest enigma is the universe's apparent razor-sharp nearness to being either "open" or "closed"—destined in the former case to expand forever and in the latter eventually to collapse in a contraction that would mimic the Big Bang but in reverse.

The rapid initial expansion, the inflation, homogenizes the universe and makes it mathematically perfectly poised between "open" and "closed"—having precisely the mass density for this condition, now about three hydrogen atoms per cubic meter of space. The universe will probably continue to expand forever but is gradually slowing its expansion rate, yet never quite to stopping. Inflation "blows away" or dilutes all those bothersome magnetic monopoles into the much larger universe and explains why we don't encounter them. And it suggests why the universe looks so uniform; any large-scale irregularities present at the first instants of time were smeared far beyond our minute domain in the tiny visible part of the much larger universe. There may be such irregularities, "walls" or other large kinks in space-time, but if so they are so fantastically remote we can't experience them.

For a long time cosmologists have known that they had a "missing mass" problem if the universe was to be closed, or at least on the knife edge between closed and open. Calculations show that the luminous mass in the universe constitutes but 1 percent or at most 10 percent of the required mass for closure! They had also inferred enough dark matter in clusters of galaxies and in the dark halos of individual galaxies that excercises an apparent gravitational influence sufficient to make these structures behave dynamically the way they do. Still, this additional mass would account perhaps for only 10 to 20 percent of the mass density required to close the universe. The remaining 80 or 90 percent of the universe's mass necessary for the balancing act may exist in some other form.

Inflation provides a theoretical reason for believing that a pervasive kind of "dark matter"—the missing mass—must permeate all of space. Regardless of the initial conditions before inflation, the theory mandates that the universe will be driven inexorably to the borderline between open and closed. For practical purposes the universe is open and will expand forever, even as it slows toward zero expansion rate at an infinite time. Cosmologist Marc Davis at the University of California at Berkeley explains that Guth's inflation would "dynamically drive the universe to the critical state exactly between open and closed—with extremely high precision. If you want to believe in inflationary theory—we've got to find the missing mass."

Dark matter, the missing mass, has gravitational influence, but it apparently doesn't react with ordinary matter in any other way. The missing mass can't be made of *baryons*—ordinary particles like protons or neutrons—because if it were, according to calculations from conditions prevailing during the early universe, the cosmic abundance of light elements like helium, deuterium, and lithium would be different from what is observed. The strange dark matter is among us and through us and we don't even feel it. One candidate for the missing mass—"mini-black holes," each with the mass of a mountain yet being smaller than an atomic nucleus—may have been created long ago when the universe was highly compressed. Perhaps ghostly, highly noninteracting neutrinos that pervade space have a small mass after all, rather than none whatsoever. A very small mass for these spritely particles could easily account for the apparent matter deficit. Some theorists offer as candidates for the dark matter less familiar, bizarrely named particles predicted by new theories of the microcosm. Whatever dark matter turns out to be, it is as though the billions of galaxies were nothing but a scanty, glowing froth in a dark, mostly invisible universe—like whitecaps on foamy ocean waves topping the far deeper sea.

Cosmic Genesis

With inflation, physicists like Alan Guth describe the universe as being "a free lunch"—created out of probabilistic fluctuations in an original *false vacuum*. In general terms, the gravitational energy of the universe exactly cancels its nongravitational energy—the two energies having opposite signs. The net energy content of this kind of universe could be exactly zero! The mind whirls in attempting to imagine how everything could have sprung forth from literally nothing—nothing but a pre-existing fluctuating "false vacuum" governed by nature's ultimately unified force law. If recent theory is on the right course, it seems that the universe is nothing but a nothingness that has elaborated itself. A formless void incubated all being. Can we conceive of this nothingness in any way? Probably not. The mind simply refuses to transcend spacetime to reach perfect nonexistence.

The conditions reigning before the first one-hundredth second of the universe would have been ideal circumstances for particle theorists to test their ideas, though of course they can never revisit those hellish times, only examine the consequences that have followed. At the Planck moment the temperature of matter was 10^{32}°K and its density 10^{96} times that of water in a kitchen sink. Particle energies amidst the staggering high temperatures were then much greater than the most fearsome nuclear accelerator on Earth is ever likely to conjure up. So to reconstruct the beauty of cosmic dawn, physicists must content themselves with available data and with theorizing.

After the Planck era, the hot, dense universe was expanding and cooling. It was an unimaginable maelstrom of radiation, heavy particles with the quizzical name *quarks*, and more fleeting ones called *leptons*. Its condition has been likened to a state of nearly *perfect symmetry* to denote an era in which, excluding gravity, three of the later emerging and differentiated forces of nature were manifested as one force. Then beginning at 10^{-36} seconds, a primordial "field" like a formless fluid throughout space, the so-called Higgs field, figuratively shrugged. The field, a pre-existing state known mystically as the "false vacuum," fluctuated and ignited the inflationary phase of the universe.

After the initial slow expansion had cooled the universe to 10^{27}°K, the temperature dropped precipitously, though by any standards it was still extremely oppressive—10^{22}°K. Then something analogous to a sudden crystallization of ultracool water to ice occurred. Heat was liberated just as when supercooled water freezes suddenly to ice, producing a momentary temperature rise. This broke the symmetry of the prevailing fields, and the so-called strong nuclear force differentiated from what we know as the electroweak force, itself the unification of the electromagnetic force and the weak nuclear force. The universe had inflated enormously by 10^{-32} seconds, and the expansion rate now resumed the "plodding" steady pace of the earlier era. Cooling was still continuing, and by 10^{-11} seconds, when the temperature was by earlier standards a frigid 10^{15}°K, the electromagnetic and the weak forces became distinct.

Physicists are driven to understand how four fundamental forces of nature are related to one another, and how these are expressed in an Alice-in-Wonderland realm of subatomic particles. Attempts to build a key segment of an expanded theoretical edifice (another part of which was once sought by Einstein as a "unified field theory") are generically known as Grand Unified Theories or GUTs—the linking of the strong and weak nuclear forces with the electromagnetic force. GUTs are at the core of what the universe was like at the beginning, and they help to predict what we should expect to find in the cosmos today. More comprehensive theories than GUTs have emerged to link gravity with the other three forces at a time earlier still than the inflation era.

The four basic forces in nature are: (1) gravity, which governs the interaction of large masses and controls the large-scale structure of the cosmos today; (2) the electromagnetic force, which rules the clouds of electrons that flit around atomic nuclei; (3) the so-called "weak" force, which controls radioactive decay; and (4) the strong force, which holds quarks together inside protons and neutrons—the garden-variety particles of the atomic nucleus. These forces were not always seen to have much in common, but there was enough mystery about their relationship for a lot of speculating. Discovery of the massive W and Z particles in 1983 confirmed the theory, proposed by physicists Steven Weinberg, Abdus Salam, and Sheldon Glashow in 1969, that the electromagnetic force and weak nuclear force are truly different manifestations of the same entity—the electroweak force. Hence, experimental evidence exists for the commonality of at least two of nature's prime operatives.

Even if laboratory evidence cannot be developed to show the unification of the strong force with the electroweak, the largest experiment of all—the early universe itself—may prove to be the ultimate test of GUTs that unify three of the fundamental forces. This is where cosmology and particle physics have joined to provide answers to common puzzles.

We leap safely into the future away from the puzzling Planck era and the symmetry-shattering events in the universe's first one-hundreth second of existence. At 0.01 seconds the maelstrom of subatomic particles and radiation swirled at a temperature of 10^{11}°K, and with a density four billion times that of water. It was the time before which matter and antimatter may have been present in almost equal amounts, frantically annihilating each other and just as soon rematerializing. But a minute asymmetry between their quantities—only about one part per billion, created during the GUT era—allowed a tiny residue of matter to survive the mutual annihilation of all the rest into pure radiation. The visible universe mercifully survives as the detritus of the matter-versus-antimatter conflict.

In the one-hundreth-second era roughly as many protons (positively charged particles) as neutrons (no charge) existed. Contrariwise, photons of radiation outnumbered the massive particles by a billion to one, a ratio that persists to our own day. Particles and photons were colliding with one another in a state of thermal equilibrium—each entity having the same energy. The universe continued to expand as the temperature dropped to 10^{10}°K shortly after one second. The density of the universe was now a comparatively tenuous 400,000 times that of water. At this time, ghostly particles called neutrinos, which may or may not have any mass, stopped interacting strongly with the other particles and became an independent ethereal background to the rest of matter in the universe. Today a vast flux of neutrinos can penetrate the body of Earth with few ever being stopped. Neutrinos continue to play a role in cosmological speculation, since if these ghosts of the past possess any mass at all they could conceivably comprise the bulk of material in the universe—the "dark matter." Like photons of radiation, they still outnumber heavy nuclear particles a billion to one.

At a quarter of a minute into the Big Bang, the temperature had dropped to 3×10^{9}°K, and nuclei of the element helium were just beginning to form. As nuclear reactions progressed, the ratio of protons to neutrons was changing from one-to-one to favor protons. When the cosmic clock struck four minutes, the remaining neutrons had been cooked into helium. The relative abundance of helium and hydrogen was established—an enduring feature of the present universe, modified only by further thermonuclear baking of hydrogen to helium in stellar interiors. Steven Weinberg, who has explored the physics of the young cosmos, nonetheless admits that "I cannot deny a feeling of unreality in writing about the first three minutes as if we really knew what we are talking about."

Until several hundred thousand years of the Big Bang had passed, radiation coexisted with matter, but the temperature was too high (more than 3,000° K) for the clouds of matter to be transparent to the radiation. Finally sufficient cooling occurred due to expansion so that buzzing clouds of negatively charged electrons could combine with light atomic nuclei to form true atoms. The universe lost its opacity to light; it came out of the primeval murk; and the stage was set for the evolution of all the structure we observe today—galaxies, stars, planets, and even cosmologists who hope to figure it all out. The once thoughtless void now thinks.

If the universe should have more than the exact critical density of gravitating matter that inflationary theories say it must have, it is "closed" and will collapse many billions of years hence, all vestiges of structure disappearing as it pops out of existence like a burst bubble. Some theorists have suggested that it would then "bounce" and re-expand in a new Big Bang. The present incarnation of the cosmos might only be one of many of an infinity of "prior"* universes which might or might not resemble our own. But a greater number of cosmologists believe that the observable universe will go on expanding forever, even as gravitation inexorably slows the stretching of space-time. Not only is gravity critical to the ultimate fate of the universe, it is central to understanding how all the forces of nature may have been aspects of a single beautifully symmetric universal law before the Planck moment, the apparent gate of oblivion at 10^{-43} seconds.

---

*"Before" the beginning of time really has no intuitive meaning.

3

THE RAINBOW AT THE END OF GRAVITY

There is a deep compulsion to believe the idea that the entire universe, including all the apparently concrete matter that assails our senses, is in reality only a frolic of convoluted nothingness, that in the end the world will turn out to be a sculpture of pure emptiness, a self-organized void.

—PAUL DAVIES, *Superforce,* 1984

The fact that the universe is governed by simple natural laws is remarkable, profound and on the face of it absurd. How can the vast variety in nature, the multitude of things and processes all be subject to a few simple, universal laws?

—HEINZ PAGELS, *Perfect Symmetry,* 1985

To ancient Greek philosophers, falling toward the ground was a natural property of all objects, violated only by the heavenly spirit, fire, which goes up, and stars, which move up and down in apparent circles. For more than 2,000 years this single statement was the sum of human knowledge about the force we call gravity. Prescientific though it was, the Greek vision was an admirable effort to perceive order in an otherwise magical universe. Galileo and Newton, the intellectual giants of the sixteenth and seventeenth centuries, took the first halting steps toward a deeper understanding of the force that so obviously governs daily life. With an unprecedented leap of

scientific imagination, Albert Einstein in the early years of the twentieth century soared far beyond his predecessors to expose gravity as a glorious manifest geometry. According to his theory of General Relativity, stones fall and planets move because of the subtle curvature of four-dimensional space-time.

Mathematics is often prelude to scientific theories, foreshadowing them in a most eerie way decades or centuries before scientific insight occurs. One of the most remarkable cases of mathematical prescience was the work of nineteenth-century German mathematician Georg Riemann, whose work anticipated General Relativity. Riemann developed a systematic study of geometries of higher-dimensional space that was intended to be only an intellectual exercise—an attempt to purify the postulates of everyday Euclidean geometry, mathematics that had been known for more than two thousand years. One variety of Riemannian geometry has the peculiar properties that infinitely long lines do not exist and through some two points an infinite number of "straight" lines may pass. This is like the mathematics of the surface of a sphere.

Quite unexpectedly Einstein's exploration of gravity showed that a Riemannian geometry applies to the universe if the cosmos is thought of as a continuum of four-dimensional *space-time* rather than the three-dimensional world of everyday experience. This was about fifty years after Riemann did his work! General Relativity says that gravitation is simply a characteristic of the curvature of space-time. Nor was Einstein speaking about the obvious three-dimensional curvature of an object like a sphere; rather he was suggesting that three-dimensional space itself could be curved within a four-dimensional realm. Matter causes space-time to curve and is in turn affected by curved space-time.

General Relativity's breathtaking unification of space, time, and matter to explain gravity nonetheless failed to integrate gravitation with the three other fundamental forces of nature (the electromagnetic force, the strong nuclear force, and the weak nuclear force). Gravity appeared always the odd force out in physics, phenomenally weaker than other forces, 10^{-39} the strength of the electromagnetic force, yet strong enough over large distances to be the organizing power in the macroscopic world of planets, stars, and galaxies. In the microcosm of atoms and subatomic particles, gravity played no perceptible role. The laws of quantum mechanics, another revolutionary insight born of the twentieth century, were required to understand the behavior of the players in the microcosm, but quantum mechanics seemed to bear no relation to gravity.

Today physical theory is undergoing a revolution that may soon tie together all the forces of nature—at last the "holy grail" unified field theory once sought by Einstein. Physicists have already unified some of the fundamental forces in the last two decades, and it appears that the last force to fall into place will be gravity. Theoreticians are hard at work on a vision of "quantum gravity," known also as "supergravity" or "supersymmetry," which if successful may be nothing less than a complete theory of the known forces of nature.

All the mysterious and evanescent subatomic particles may one day live in the framework of a unified field theory, interacting with one another according to well-defined laws. And this edifice will represent more than mere theoretical beauty. At sufficiently high temperature (10^{32}°K), such as prevailed during the Planck era at the start of the universe, all forces are expected to blend together and to act as one. In these theories, on an unimaginably small scale the very fabric of space-time acquires a complex structure far beyond our normal apprehension of space as uniform emptiness. Physicists working on supergravity speak seriously of an eleven-dimensional universe, a dimensionality required to encompass all the known particles. We may be unaware of the other dimensions in daily life only because they lurk tiny and curled up, hiding in the subnuclear world.

Cosmologist Paul Davies applies the aptly expansive term "superforce" to the unified natural law. In his book *Superforce* he boldly suggests that "for the first time in history we have within our grasp a complete scientific theory of the whole universe in which no physical object or system lies outside a small set of scientific principles." Summarizing the new vision of physical law, Davies states, "The world, it seems, can be built more or less out of structured nothingness. Force and matter are manifestations of space and time. If true, it is a connection of deepest significance."

Back in the seventeenth century gravity was the first force of nature to come under scientific scrutiny. The study of gravity and the motion of bodies under its influence, in effect, ignited the scientific and technological age. How inspiring that these early studies appear to be culminating now in an all-encompassing theory of existence! But the road to success wasn't easy. Galileo required the courage to challenge the Aristotelian view that heavier bodies fall faster than lighter ones. In establishing Aristotle's error and proving that all bodies accelerate toward the ground in the same way, Galileo started a path that Newton and then Einstein pursued with a vengeance.

In its key "principle of equivalence," General Relativity enshrines the basic concept that acceleration caused by gravity and acceleration caused by other forces are fundamentally indistinguishable. This means that the quantity of mass used to compute force due to gravity, *gravitational mass*, and the quantity of mass causing the inertia of that same body, *inertial mass* (its resistance to acceleration), are identical. In experiments with falling objects, Galileo, in effect, proved this equality. The principle of equivalence is the reason that an astronaut can hover motionless near a 100-ton spacecraft while falling around the world in orbit.

Isaac Newton, a sickly ninth child, was born in 1642, the year that Galileo died. When Newton was a twenty-three-year-old college student, he escaped to the countryside to avoid the bubonic plague. In that interlude he formulated his law of universal gravitation, which describes equally well the fall of apples and the fall of the Moon toward Earth. What an astonishing prospect, to be able to characterize the motion of heavenly bodies with rules governing the fall of a sparrow! Newton also proved that the evident ellipticity in Mars's orbit around the Sun, the monumental discovery of Johannes Kepler and Tycho Brahe, followed from his "inverse-square" law of gravity. As high school students learn, the force between two bodies decreases as the square (self-multiplication) of their distance of separation is divided into the product of the bodies' masses. Meticulous Newton developed an entirely new branch of mathematics—calculus—to prove the idea, which seems intuitively clear enough, that the sum the gravity forces produced by every particle of matter in the Earth tugging on a small body is equivalent to the entire mass of the planet acting as though it were concentrated at the center of the world.

Newtonian gravity is elegant simplicity. Yet for centuries it sufficed to describe accurately the course of planets and their moons, the tides, the wobbling of Earth like a top (one wobble every 26,000 years), and the orbital fall of a satellite around the globe—a possibility foreseen by Newton himself! In the nineteenth century the laws of electromagnetic force and electromagnetic radiation (light, radio waves, etc.) were worked out by scientific luminaries such as Faraday, Maxwell, and others. At the end of the century many physicists were confident that all the fundamental laws had been discovered. All that remained was to apply the laws of gravity and electromagnetism in ever more sophisticated ways to explain the complex phenomena of the natural world.

Physics was soon shaken by cracks in its theories that became gaping chasms. Electromagnetic waves (light) were thought to propagate through an invisible fluid medium called the aether. What a shock to find that the speed of light was constant for all observers, no matter how fast they were traveling toward the source of light! This is far from an obvious circumstance, for in everyday life we are accustomed to the addition of velocities—a baseball thrown from a moving car strikes a roadside wall with the velocity of the throw added to the automobile's velocity.

Einstein resolved the seeming paradox of light's constancy by dispensing with the aether and formulating a new intimate relation between space and time—his theory of Special Relativity. The theory extended Galileo's idea of mechanics that an observer could not detect his motion if he was moving at constant velocity. With Einstein's extension of that philosophy, even experiments with light inside an observer's compartment would not reveal his motion. Gone was the popular notion of a universal time called now. The tide of time was inextricably connected with space, and events occurred in space-time rather than in space alone.

At the turn of the century accumulating contradictory evidence dealt mortal blows to the idea of light as a pure wave phenomenon. The atom, no longer the hard billiard ball of classical physics, was beginning to show structure—an electrically charged nucleus with surrounding electrons. That an atom had a tiny nucleus was not known until experiments and analyses conducted from 1906 through 1911 by physicist Ernest Rutherford and others. A new theory, quantum mechanics, then had to be developed to explain atomic phenomena. In quantum mechanics particles and light have dual attributes of particle and wave, and they behave according to the laws of statistics rather than deterministically. The microcosm acquired a capriciousness that continues to shake our world view. Einstein, though a major contributor to the development of quantum mechanics, refused to believe that "God plays dice with the universe." Today most physicists believe that although God may not play dice, Nature surely does. Subatomic particles make transitions in their states of motion, position, spin, etc., that are impossible to predict and that are unlike any common behavior in the macroscopic world. The probability of occurrence of a particle's transition can be computed with great precision, but the occurrence itself cannot be predicted based on "hidden clockwork."

With this as background, gravitational theory took a giant leap with the publication of Einstein's theory of General Relativity in 1916, which could be considered an attempt to extend the ideas of Special Relativity to situations involving acceleration, not simply uniform motion at constant velocity. Would an observer in an enclosed chamber be able to determine whether rocket propulsion or gravity was pushing him to the floor? In principle it would be impossible to tell, hence the "principle of equivalence" between gravitational and inertial accelerations. Einstein's leap of imagination was to suggest that not even beams of light inside the chamber could discriminate gravitational acceleration from acceleration caused by contact forces. The elegant theory that follows from this assumption makes predictions, and it was not long before some of them were confirmed.

In 1919 a solar eclipse observed from Africa showed that starlight was indeed bent by the Sun's gravity—a tiny amount just as predicted by Einstein. Newtonian physics also predicted a slight bending, but not so great as the one observed. The planet Mercury's closest approach to the Sun, its perihelion, had mysteriously rotated 43 seconds of arc per century (as seen from the Sun) more than could be accounted for by gravity from the other planets. Again, General Relativity predicted this amount of shift. Generally, the strength of a gravitational field has to be great before deviations from Newton's law of gravity are noticed. This is why Newtonian gravity was considered to be the final word for so long. General Relativity also anticipated that time should slow in a gravity field, and this too has been confirmed. A heat and pressure resistant watch ticking for a year on the surface of the Sun would lose about 30 seconds compared to a terrestrial clock.

But successful predictions alone do not measure the worth of a physical theory. It has most value if it can encompass a variety of phenomena, forging links with other physical laws. General Relativity was eminently successful in connecting gravity with the curved geometry of space-time. According to the theory, masses moved on trajectories—*geodesics*—in four-dimensional space-time that were simply the straightest possible path between points for that curved geometry. This conceptual elegance expressed in geometrical terms was quite different from the mechanisms of the other fundamental forces. The pioneering theory of *quantum electrodynamics*, for example, had been developed in the 1930s and was codified in 1948 and 1949 to explain the electromagnetic force between charged particles as an exchange of fleeting *virtual* particles—in this case photons ("particles" of light). The term "virtual" applies, because they scoot in and out of existence before any observer has a chance to detect them and thereby violate fundamental quantum mechanical principles.*

The strong nuclear force, responsible for holding the nucleus together, and the weak force, that governs the radioactive decay of particles, were also explained as the exchange of evanescent "messenger" particles. Like the photon intermediaries of electromagnetism, the exchange particles were of a general class called *bosons*. Exchange particles appeared to explain electromagnetic force and the two nuclear forces, but no corresponding exchange particle for gravity was in sight.

Physicists look for symmetries in their attempt to find new physical theories. The symmetries often have the property that under a certain type of mathematical transformation, physical laws retain their original form—remain invariant. In order to create symmetry, physicists are led by mathematical considerations to introduce new fields from which new forces

---

*Virtual particles are a consequence of the Heisenberg uncertainty principle of quantum mechanics. If the uncertainty in time is small enough, then uncertainty in energy is large enough to create massive virtual particles from the vacuum.

may appear. These fields, called *gauge fields*, are associated with particles of exchange, *gauge particles*, that shuttle undetected between interacting parent particles to produce force.

One such symmetry, discovered in the 1960s, dramatically unified the electromagnetic force and weak nuclear force into a common electroweak force. The exchange or messenger particles of the weak force (W and Z particles) were discovered at the powerful European CERN nuclear accelerator in 1983. It was now apparent that before the first 10^{-11} second of time, when the temperature was higher than 10^{15}°K, the electromagnetic and the weak force were one, and shared the same exchange particles. When the expansion of the universe caused the temperature to drop below this level, the two forces differentiated due to "symmetry breaking."

There has also been a conjectured unification at a higher temperature of the strong nuclear force with the electroweak force in grand unified theories (GUTs). A consequence of this supposed unification is that all protons in the universe will ultimately decay. Physicists have been working hard to detect the extremely rare decay of a proton, and some believe that evidence has been found for this. The force unification brought about by GUTs occurred earlier than the first 10^{-31} second of time.

Could gravity have been unified with the other now conjoined forces at a still earlier time—the time of the "superforce"? Theorists have been speculating about this since the early 1970s to find the elusive theory of quantum gravity and cast it in a comprehensive theoretical framework. At least eight possible supersymmetry theories are known that might connect all the myriad particles that inhabit the universe. But the eighth-level theory, called arcanely SO(8) supergravity, was a promising possibility in the late 1970s. However, even this, the supergravity theory that has the most particles, specifies too few of them to agree with the physical world. And SO(8) remains plagued by a host of mathematical difficulties.

Physicists will have to resolve inconsistencies between these theories and observation before they can say that they have connected all forces and particles.

Physicists designate the exchange particles for supergravity the "graviton" and the "gravitino." One beautiful aspect of supergravity is that General Relativity can be derived from the shuttling of gravitons and gravitinos between particles. A theory that once consisted of purely geometrical relations can now be interpreted on another level as particle interactions. Yet the pendulum of physics is swinging again, bringing in new geometries of higher dimensional space to explain the particle interactions of supergravity. In particular, a five-dimensional space-time geometry, developed by Kaluza and Klein in the 1920s to tie gravity and electromagnetism together, was resurrected and expanded to 11 dimensions as part of an effort to create a *dimensionally extended supergravity* theory.

By adding seven space dimensions to four-dimensional space-time, physicists have been able to describe their menagerie of particles as the curling of space-time at incredibly small scales. We are unaware of these added dimensions except for their manifestation as particles. According to extended supergravity theory, down at the so-called Planck distance, 10^{21} times smaller than an atomic nucleus, space has a foamlike structure. Gravitational theorist Bryce DeWitt writes that ". . . an observer who attempts to penetrate the fourth spatial dimension is almost instantly back where he started. Indeed, it is meaningless to speak of such an attempt, because the very atoms of which the observer is composed are larger than the cylindrical circumference [of the fourth dimension]. The fourth dimension is simply unobservable as such."

Our conceptions about space and time are sorely tested by the framework of extended supergravity. Tiny wormholes, tears, and lumps may appear in a statistically fluctuating fabric of space-time at the minute level of the Planck distance, even in the present era, not simply when the universe began at the

Planck moment. Bryce DeWitt writes, "In a universe governed by quantum gravity, the curvature of space-time and even its very structure would be subject to fluctuations. Indeed, it is possible that the sequence of events in the world and the meaning of past and future would be susceptible to change." Paul Davies writes that "The orderly arrangement of points, the smooth continuity of the space of classical geometry disappears in the froth of space-time. Instead we have a melee of half-existing ghost spaces all jumbled together. In this chaotic shifting sea, the common sense notion of 'place' fades completely away."

If extended supergravity seems remarkable to the nonspecialist, it has also enthralled many physicists. Previously, there stood two major classes of subatomic particles, bosons and fermions. The groups were distinguished by what physicists call *spin*—bosons having only integer values of spin and fermions having half-integral values (½, 3/2, etc.) Two different fermions cannot occupy the same state. Thus, atoms have extended shells of electrons (which are fermions) because of this inherent fermion property. Without this *Pauli exclusion principle* of quantum mechanics that prevents the electron clouds of atoms from collapsing, the universe we know could not exist. By contrast, bosons can occupy the same state—a beam of laser light having many overlapping photons is a good example (photons are bosons). No one suspected that these two radically different classes of particle might be tied together by an underlying symmetry rule. Fermions had been associated with matter and bosons with force. *Supersymmetry* connects the two classes in a completely unsuspected way. Supergravity is the *force* that preserves supersymmetry among particles in the mathematical sense required by physics.

In recent years a controversial and rapidly expanding field of theorizing has opened fantastic new vistas to help explain the unity of forces and the nature of particles. Physicists call the new ideas superstring theories, which postulate that elementary particles aren't infinitesimal points after all, but tiny one-dimensional *strings* that constitute the most fundamental geometry of space. And these figments of existence also apparently have masses. The tiny vibrating stringlets may be either little lines or loops, and their lengths are of the order of the Planck dimension—10^{-33} centimeters. Their characteristic vibrations (the ultimate "music" of the cosmos?) determine which subatomic particle each represents.

Superstring theories require a ten-dimensional space—nine space dimensions and one time dimension. Six of the space dimensions are somehow curled up and hiding from human perception deep within the Planck microcosm, though theorists suggest that at the birth of the universe all the space dimensions were equally manifest. Superstring theories nicely seem to encompass supersymmetry theories, and they are better than other ideas in that they explain the mass ratios among fundamental particles. Controversy abounds, however, because beautiful and consistent as superstring theories are, no one has the faintest idea how any experiment will ever be able to establish them. The required energies are beyond any that civilization is likely to master—10^{16} times our present capability.

Gravity was once the dusty curiosity of high school physics labs with their inclined planes, spring scales, and stopwatches. It now appears that science has come full swing to view gravity—blended together with the other basic forces, to be sure—in some sense as the central creative agency in the universe. We knew that gravity stoked the fusion fires of stars, that it made planets and moons round, and that it plotted the course of galaxies. We did not suspect that an enlarged view of gravity—the superforce of supersymmetry and now superstring theory—would be the key perhaps to all physics and the origin of all things. The end of gravity would be planets that go around stars, physicists postulating geometry, and people chasing rainbows. The end of gravity was to be the rainbow spectrum of life itself.

4

THE ANTHROPIC UNIVERSE

In broad terms, the history of the universe is the story of the mountain that gave birth to a mouse. . . .

Was all of the unfolding of the universe in space and time written out in the play of interactions between particles? The flight of a swallow, Beethoven's last sonatas, or your next weekend at the beach—were they already composed in parts that quarks, electrons, and photons prepared for reading and performance 15 billion years ago?

—HUBERT REEVES, *Atoms of Silence,* 1984

Because of our woeful ignorance, everything fundamental seems to us to be either accidental or designed.

—EDWARD HARRISON, *Masks of the Universe,* 1985

In brief, how can the machinery of the universe ever be imagined to get set up at the very beginning so as to produce man now? Impossible! Or impossible unless somehow—preposterous idea—meaning itself powers creation. But how? Is that what the quantum is all about?

—JOHN ARCHIBALD WHEELER, 1986

The universe, born of absolute nothingness, the fulfillment of a geometry that transcends time and space, is now germinating in variegated forms barely on the first morning of cosmic evolution. Can it all be just a happy conspiracy of coincidences emerging from lawless chaos?

Going against centuries of scientific tradition that has been neutral or indifferent to such questions, some cosmologists—voices in a wilderness of doubt—have proposed that the universe has been perfectly "designed" to accommodate life in a way that could not have happened "by chance." Attributing "purpose" to existence—a teleological "destiny"—is guaranteed to raise heated opposition in the scientific community, and it has. Still, prominent investigators express these sentiments that are so peculiar in the traditions of science. British cosmologist Paul Davies, in his ironically titled book *The Accidental Universe*, writes, "Recent discoveries about the primeval cosmos oblige us to accept that the expanding universe has been set up in its motion with a cooperation of astonishing precision." Some theorists, such as quantum physicist John Archibald Wheeler, have gone so far as to suggest that if the cosmos never evolved a phenomenon like life, that observes and contemplates, it would be a pointless if not impossible universe. Yet they are far from claiming what conventionally might be thought of as a "proof of God." Rather, they are exploring the very meaning of "chance creation." What a surprise, to see the usual domain of religion invaded by mainstream science, causing discomfiture among many scientists and no doubt religionists too.

Scientists believe that life, as we know it on Earth, originated and evolved on a planetary surface only by the grace of many congenial circumstances—not too warm, not too cold, the right chemicals, the right energies, neither too little nor too much stability in the environment. It is simply common sense that any one of a number of small or large changes could prevent or substantially alter the course of life on a planet. Witness, for example, the demise of the dinosaurs, an event perhaps caused by a colliding asteroid or bombardment of comets, that dramatically changed the course of evolution.

But these fortuitous details of biological evolution, fascinating though they may be, are not what intrigue cosmologists.

They question instead why fundamental physical conditions allow atoms to exist, permit stars to form, and let the universe's bubble of space-time expand without destroying the magnificent pattern of galaxies within. In short, why the physical laws are as they are with their particular forms and numerical constants. It is extraordinary that science has come so far that it can question the "why" and not just the "what and how" of physics. In a sense, for modern cosmology, "how" is becoming almost indistinguishable from "why," much as the content of certain branches of mathematics is becoming almost identical with the most fundamental physics.

The Copernican revolution began in the fifteenth century and displaced the Earth as the center of the universe. The Sun, not the Earth, became the center of the solar system. Then the Sun was demoted to but one of many other stars, appearing pointlike and distant in space. By tedious investigation, the stars were later organized into an island universe of hundreds of billions of stars—the Milky Way galaxy. And finally the Milky Way was itself dethroned and became only one of hundreds of billions of galaxies scattered through space. Now the inflationary universe appears to expand our horizons far beyond the already fantastic vista of those billions of galaxies. What next?

This process of deprovincialization has been very sobering and beneficial to the unchecked human ego, but it has had the unfortunate side effect of seeming to demote the importance of life in the universe. Even if life existed on other worlds, what significance could it have compared to the vastness of time and space? If it was just an "accident" that life arose on a particular planet and acquired a certain form, surely it could have no more compelling rationale than the equally "accidental" major features of the universe—galaxies, stars, and planets. Now the pendulum is swinging in the opposite direction, and some cosmologists are asking whether life—organized, contemplating complexity—may be a phenomenon of central importance in the cosmos.

The problem of asking why the physical laws underlying the cosmos are as they are is that, by definition, we have only one example of a universe, with one set of laws presumed to be governing it. As with the mystery of life's origin, there is but one example to study. There is hope and expectation that we will find other examples of life on other worlds, independently arisen. But can we find another universe?

If we can't really find another universe, cosmologists have opened wide the door of possibility that there may have been "before" the present incarnation of existence other universes. These and other co-existing universe domains are perhaps forever beyond our physical reach, yet may be touchable by theory. Parallel or antecedent worlds could have had laws either slightly or radically different from our own. The immediate natural question: why is our universe so "nice"—so conducive to sentient life with such faltering aspirations of self-knowledge? This often prompts the sarcastic response that reduces the issue to circular reasoning or tautology: "We wouldn't be here to ask that question if we weren't here." Or even more mocking: "The universe exists because it exists." Some but not all physicists were satisfied with those responses to an apparently blind-alley question.

Codifying all the ways in which the universe is congenial to life has been a great challenge to cosmologists who have tackled the problem. Fortunately, there is now a fairly well-established consensus on an overall description of the universe. The beginning occurred about 15 billion years ago. The universe came out of the Big Bang expansion of all space-time starting then. All matter didn't explode from a highly compressed state into an infinite void. Rather, space-time itself blew up from a virtually infinitesimal point, continuing to expand and allowing galaxies of stars to evolve and grow ever farther apart. Moreover, this point is infinitesimal in the almost unimaginable sense of a spherical surface that shrinks ever smaller toward a point, or an infinite volume that squeezes down to oblivion in an even less intuitive way.

We have encountered inflationary theories that characterize the first instants of creation and that have helped to explain some of the thorniest problems bedeviling cosmology. Above all, inflation seems to sanction the idea of the universe as a "free lunch"—all of matter being created from literally nothing, from the bizarre pre-existing manifold described as a fluctuating "false vacuum." And attending the birth of the universe in its first millionth of a second, we have remarked about the extraordinary separation into distinct forces of what was once a fundamentally unified force of nature, the "superforce." The superforce differentiated at that time into the four fundamental forces that now govern the universe.

Many believe that the first physicist to hint of something strangely coincidental about the cosmos was the late Nobel laureate and pioneer of quantum mechanics, Paul Dirac, who did so in 1937. But physicist John Wheeler recounts that German physicist Hermann Weyl anticipated the thrust of Dirac's reasoning as long ago as 1919. Influenced by the earlier suggestions of Weyl and others, Dirac noted that numbers with the "order of magnitude" 10^{40} kept appearing unexpectedly. This was the approximate ratio of the huge electromagnetic force between two particles each with unit charge (for example, protons or electrons) to the much weaker gravitational force between them. Dirac also noticed that the age of the universe ratioed to the time for light to zip across a proton's length (the most fundamental distance he knew about) was also about 10^{40}. The number of protons that would equal the mass of the visible universe was remarkably also that magic number multiplied by itself, or about 10^{80}. These were all "pure numbers" with no units like feet, meters, or hours—anyone in the universe would derive the same number, albeit written in some alien system of enumeration.

Dirac believed that this showed a "deep connection between cosmology and atomic theory"—between the macrocosm and the microcosm. According to Dirac's wildly speculative "large number hypothesis," either the strength of gravity or the mass of the universe might be changing with time. In 1961, Princeton physicist Robert Dicke challenged Dirac's hypothesis, remarking about the present age of the universe that was key to Dirac's theory, ". . . it is not a 'random choice' from a wide range of possible choices, but is limited by the criteria for the existence of physicists." We are alive now because, much earlier or much later, conditions would not give rise to and support life. This was perhaps the first time that anyone had drawn a scientific connection between cosmology and the conditions necessary for life.

British astrophysicist Brandon Carter reignited interest in this form of scientific "numerology" in the early 1970s. He began to codify a variety of cosmic conditions necessary for life under the general rubric of what he called the anthropic principle. The anthropic principle is now the center of a storm of controversy, unleashed most recently by the publication of *The Anthropic Cosmological Principle* by British physicist John Barrow and American physicist Frank Tipler. There are actually at least two versions of this principle, which have acquired respectively the names the "weak" and the "strong" anthropic principle. The weak anthropic principle merely says that since we are alive and contemplating the universe, any cosmological observations and theories had best take that into account and realize that we live of necessity under special conditions. In Brandon Carter's words, "What we can expect to observe must be restricted by the conditions necessary for our presence as observers."

The weak anthropic principle is an interesting but almost platitudinous notion. The strong anthropic principle, on the other hand, says that any meaningful or real universe *must* evolve life. In contrast to the governance of the weak principle, the universe is not simply passing though a phase in which, of course, conditions are right and we exist. Rather, the basic physical laws are seen as *requiring* life—observers—to evolve ultimately! Are we dreaming? Has the Copernican revolution run full circle to embrace the notion that the universe revolves around life? Many resist the anthropic view, but some cosmologists seem indeed to have turned completely around. They don't say, however, that they have found among the stars the handiwork of a Grand Designer who imbued the universe with a predestiny that includes life.

Physicists B.J.Carr and M.J.Rees reviewed the state of knowledge about anthropic coincidences in a 1979 article in *Nature*. After detailed mathematical examination they concluded that "Several aspects of our universe—some of which seem to be prerequisites for the evolution of any form of life—depend rather delicately on apparent 'coincidences' among the physical constants. The possibility of life as we know it . . . is in some respects remarkably sensitive to their numerical values." Paul Davies drew much the same conclusions in *The Accidental Universe*, as do Barrow and Tipler in their compendious 1986 book of elegantly detailed arguments, some of which, though, seem misguided.*

What are some of the coincidences that prompt this chorus of physicists venerating the elegant design of the cosmos? One of the most amazing facts about the universe is that it is bathed in the cold afterglow of an extremely hot creation billions of years ago—convincing proof discovered in 1965 that the universe did have a beginning. Astronomers pick up the so-called 3 degree Kelvin radiation with their radio telescopes—a hiss that is uniform in intensity in all directions to a quality better than one part in ten thousand. This uniformity in all direc-

---

*Particularly the suggestion that we may be the only civilization in the universe. Most astronomers stand by the compelling "principle of mediocrity," which says simply that given the vastness of space it is unlikely that human civilization is a unique phenomenon.

tions, called isotropy, had to be set up by astoundingly uniform initial conditions in the Big Bang. If the radiation were not highly isotropic, physical reasoning suggests the temperature of space would now be intolerably high and the universe could not support chemically organized life. Paul Davies wrote, "The present temperature of space requires that the expansion rate at [the earliest time] be fine-tuned in different directions to within one part in 10^{40}. This is another stunning example of 'cosmic conspiracy.'" Inflationary theory, we have seen, makes the apparent conspiracy less imponderable but no less remarkable. In fact, given the weak anthropic principle as a governing rule, physicists might have predicted that an explanation like cosmic inflation would be forthcoming to explain cosmic isotropy.

Gravity is incredibly weaker than the electromagnetic force that makes the dense nucleus of an atom hug a cloud of fleeting electrons. If gravity had been slightly stronger—still allowing it to be much weaker than electromagnetism—stars would have burned much faster and hotter and the Sun would long since have used up its hydrogen. Life would probably never have had a chance to evolve. Had gravity been slightly weaker, all stars would have been dim "red dwarfs." The gentleness of gravity allows processes to occur that take aeons of time.

The most ubiquitous nuclear particle in the universe is the neutrino, one billion of them for every proton and electron—another relic of the Big Bang. They hardly interact with matter and have been thought to have no mass, just like "particles" of light—photons. But recent experiments have revealed that some neutrinos may have a very tiny mass. Up to a certain point a small neutrino mass would be no problem, but even a tiny neutrino mass might cause the universe to collapse on itself due to the increased gravitational restraint on the expansion. The cosmos hasn't collapsed because either the neutrino is really massless or is providently not massive enough to "close" the universe. Some theoreticians have proved that too massive neutrinos would also have prevented the formation of galaxies. We should be thankful for small things.

Life as we know it requires the spectacular deaths of massive exploding stars called supernovae. Near the end of their lives, supernovae "cook up" many elements heavier than helium and scatter those atoms all over the universe to be recycled in the bodies of living organisms and in the bowels of planets. The strength of the weak nuclear force appears to be finely tuned so that swarms of neutrinos can do their job of blowing out the freshly cooked elements in a cataclysmic supernova explosion. The weak nuclear force is neither too strong nor too weak and thus facilitates the planet-forming, life-giving phenomenon.

Another numerical "accident": the weak nuclear force is related to the strength of gravity in such a way that hydrogen rather than helium emerged as the dominant element in the earliest stage of the universe. If that relation had been slightly different in one direction, then helium instead would have been most abundant and there would have been more rapid burning and shorter-lived stars. More important, there would have been almost no hydrogen to form water. Where, then, would life be? What looks like an accident of nature, connecting separate realms of physical law, turns out to be a sensitive balance that mandates the nuclear composition of the universe.

The strong force binds protons and neutrons together in the atomic nucleus. If that force were only one-half its value, then such essential elements as carbon and iron would have been unstable and would not have survived long. An only 2 percent increase in the nuclear force strength would have burned up all the hydrogen catastrophically early in the universe.

The list of cosmic coincidences goes on and on. The skeptic might suggest correctly that there could be an unknown underlying physical principle (other than the anthropic one) that would force the universe to be so hospitable to life in all these ways—a fact that Paul Davies would find no less remarkable. He writes that "Whether the laws of nature can force the coincidences on the universe or not, the fact that these relations are necessary for our existence is surely one of the most fascinating discoveries of modern science."

Physics already had some theories lying in wait that could explain the multitude of anthropic coincidences, but at the expense of adopting an intellectual burden equally disturbing—or delightful! One solution noted by physicist John Wheeler is that there is an "ensemble of universes," each with its own set of physical laws and constants. We just happen to be in one universe that is congenial to life. These worlds might exist in parallel—or sequentially, if it were the habit of universes to expand and then recollapse. In some of them there could be structure and forms of "life" that we could never imagine. But in the words of physicists Carr and Rees, "Most are 'stillborn,' in that the prevailing physical laws do not allow anything interesting to happen in them; only those which start off with the right constants can ever become 'aware of themselves.'" This may be a way of getting a good universe "by chance."

Then there is the "many worlds" interpretation of quantum mechanics that a number of physicists seem to find more palatable than the conventional or Copenhagen interpretation. In the Copenhagen interpretation God *does* "play dice with the universe," and quantum mechanical events occur in the microcosm without apparent rhyme or reason, obeying a statistical law that abolishes causality for microscopic events. The many worlds theory, invented by Hugh Everett in the 1950s, suggests that at each measurement made by an "observer" the universe bifurcates into a treelike infinity of parallel and disconnected worlds. All possible things happen "somewhere." In one universe Erwin Schrödinger's now famous imaginary cat who was subjected to a peculiar form of quantum mechanical roulette dies from poison, and in another he continues to live. A set of parallel universes, each perhaps expanding infinitely, means opportunity to get those anthropic "coincidences" by sheer chance.

Paul Davies expresses a feeling that many people might have about multiple universes: "To explain the coincidences by invoking an infinity of useless universes seems like carrying excess baggage to the extreme. Yet it must be concluded that the alternatives—a universe deliberately created for habitation, or one in which the very special structure is regarded as a pure miracle—are also open to philosophical challenge." The multiworld hypothesis seems to violate the traditional scientific adage called "Ockham's razor" (the invention of fourteenth-century theologian and philosopher William Ockham), which admonishes us to select the simplest of a number of competing theories. Another developing view is that only one set of physical laws—*forced* by inexorable mathematical constraints—may be able to specify a universe in which all relations and natural constants are the only ones possible. To coin an intriguing phrase, "God may not have had a choice."

Philosopher Kenneth Winkler rendered a philosopher's view of the anthropic principle, claiming to find concealed anthropocentric assumptions in the anthropic principle. He challenges the principle's proponents: "If the ultimate outcome of the history of the universe is an Earth populated by more primitive life-forms, or one devoid of all life, would the defenders of the principle be prepared to say that it was for the sake of this end that the physical constants are as they are?" We know so little about all the possible ways "life" could exist even in our own universe. Perhaps some of the strange nuclear or electromagnetic life-forms imagined by physicists turned science fiction writers should have their "anthropic" principles too. They might be able to exist under a broader range of conditions. Anthropic principles specific to carbon-based life as we know it would be in potential conflict with the conditions necessary for thoroughly alien life-forms to emerge.

John Wheeler, a firm believer in the central role of quantum mechanics in the universe, takes an even more extreme anthropic view he describes as the need for "observer participancy" to sustain the cosmos. In his view, "observers" whether inanimate event recorders, such as rocks that retain traces of cosmic rays, or conscious beings, are in a real sense needed to "create" the universe. They establish a link of observation through an unimaginably vast network of elementary quantum phenomena extending back to the beginning of the universe. It may stretch credulity to believe that in any sense we—life—helped create the universe. But then "ordinary" quantum mechanics, which good physicists believe twice a day, already tortures naïve common sense.

Some will take from this the impression that tinkering physicists have stumbled upon a Grand Designer—God—hidden within theories and equations. Others, adopting an agnostic view, may simply conclude that, whether by design, chance, or process not yet conceived, we inhabit a very good world, a very promising universe. It is apparently written in the great book of the skies.

5
Order from Chaos

But shouldn't the simple have already potentially encompassed the complex? Where were the seeds of complexity during the first minutes of the universe? . . .

We stand amazed before this frenzy of organization. Matter seems able to draw advantage from even the most adverse circumstances. . . .

The organization of the universe demands that matter abandon itself to the games of chance.

—Hubert Reeves, *Atoms of Silence*, 1984

The universe contains vastly more order than Earth-life could ever demand. All those distant galaxies, irrelevant for our existence, seem as equally well ordered as our own.

—Paul Davies, 1986

The hierarchy of anthropic principles, much disparaged by physicists of Philistine disposition, may someday be discarded as useless intellectual baggage, a passing fad and delusion of spiritually motivated scientists. Physicists like John Wheeler, who can't accept a universe that doesn't ultimately give rise to "observers," may be misguided. Perhaps inflation and the Big Bang could have happened differently, producing a uniform, featureless cosmos with no place for life. But the universe is not without a texture and bears the signature of delicate primeval chaos overlying amazing uniformity. Its largely barren

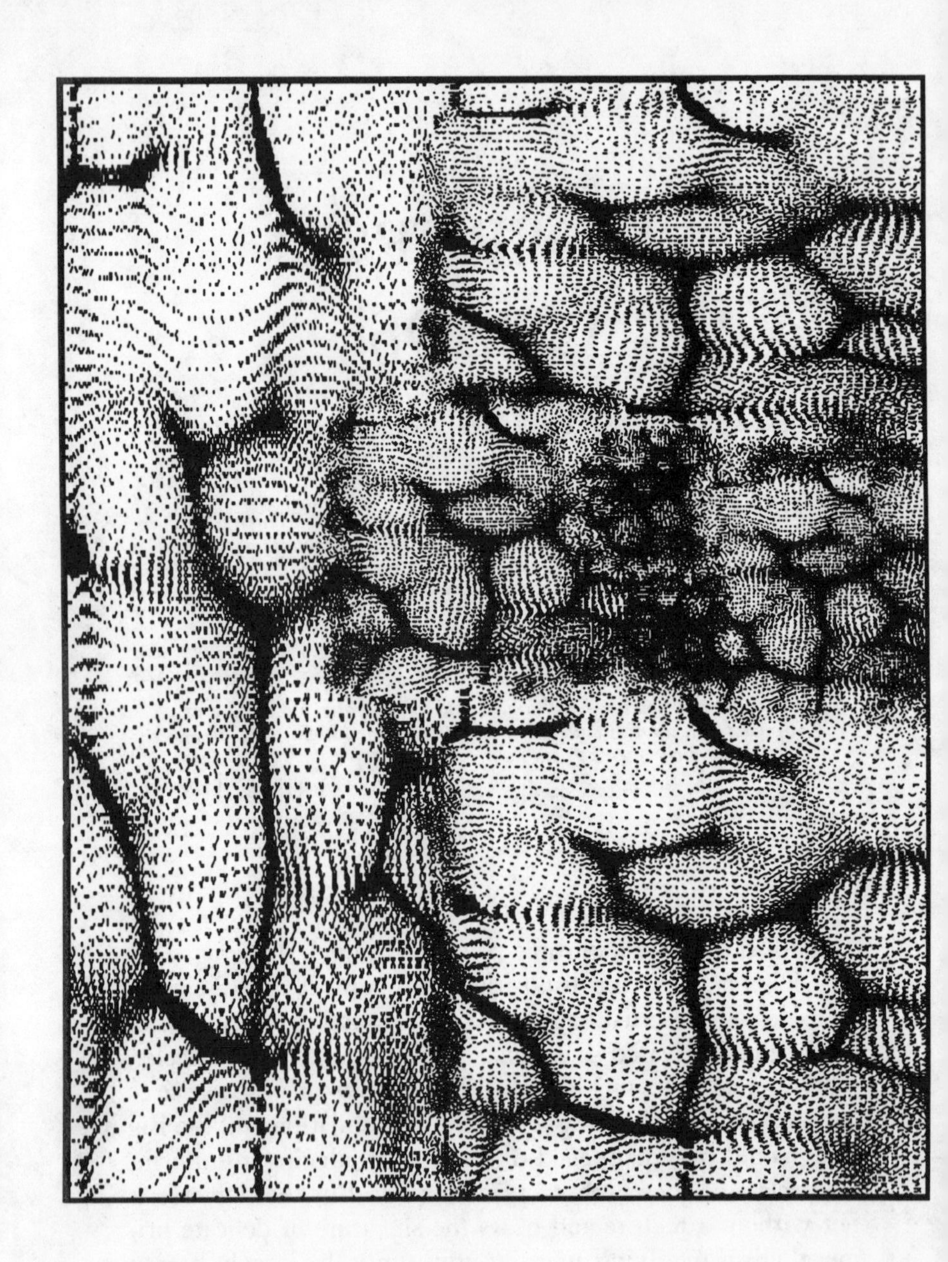

emptiness has a tenuous weave of luminous galaxies superimposed on an invisible background of gravitating substance—the unidentified *dark* matter that may comprise more than 90 percent of the cosmos. And hiding in that threadbare cosmic garment lie sites for life and intelligence, impassioned observers of it all. Above all, life needs places to live, and the universe offers them in abundance. We stare in awe at the intricacy of the pattern of galaxies, stars, and planets. How did this order emerge from chaos?

Variations on a Cosmic Theme

The seeming paradox of order emerging from chaos has been apparent only since physicists in the mid-nineteenth century stumbled across the now legendary Second Law of Thermodynamics. The amount of disorder or *entropy* of matter and radiation is by this universal legislation destined to increase toward a maximum or equilibrium value as the universe ages. The cosmos should be headed toward a "heat death" of temperature uniformity and total disorganization. How, then, does all this starry order grow and seem to flout the stern cosmic rule?

First, on the "local level" (for whatever small sub-universe "local" is defined) there is no paradox in order emerging from chaos as long as the little region is *open* so that available energy from outside can act to produce order in that small area. Almost any terrestrial life-form is an excellent example of order being produced by externally derived energy—usually from the Sun—always at the expense of more disorder in the form of heat and decay being dumped into the organism's environment. Life does not magically violate the Second Law; order increases locally but decreases universally. Yet if we accept the severe decree of the Second Law applied to the cosmos, entropy—molecular and subnuclear confusion—must be on the rise for the universe as a whole. But there must have been some initial "winding up" of the cosmos to produce the original order from which all decay proceeds.

The first seconds of the universe were a revel of disorder as primeval matter seethed in thermal confusion. Then the growing expansion of the universe imposed a remarkable order on the chaos. It is as though the universe were really an "open" thermodynamic system, analogous (but not perfectly so) to a cylinder containing gas from which a piston is withdrawn by external force. As the piston rises, the molecules of gas have freedom to move farther apart and don't collide as frequently. Likewise, by expanding and cooling, the universe permits intricacy to emerge and to be undisturbed for a while in localities. So the expansion of the universe itself acts as the agent of order, percolating organization downward through galaxies, stars, and life. The gravitational potential of the expanding universe, in effect, acts as an enormous reservoir of order—negative entropy.

The order wrought by that titanic expansion is at this point expressed in glowing stars, relics of creation that consume the hydrogen fuel "frozen out" in the first few minutes of the universe. By shining brightly, a star blasts the surrounding space with thermal disorder, yet clever species on cool worlds, finding their place in the sun, ring out a consummate order as energy slides downhill to thermal oblivion. It could have been otherwise. The expansion, if not sufficiently energetic, might have cooked up atoms of iron in the cosmic nuclear furnace, leading to myriad dead iron suns. Bright stars burning light elements are evidence of the immense *disequilibrium* or ordering engendered by the cosmos, an organization that is inexorably decaying toward an uncertain rendezvous with the arrow of time.

The Veil of the Heavens

The universe is nothing if not hierarchical; wheels within wheels, organizations within organizations are its essential flair. Quarks form nucleons, nucleons marry electrons and build atoms, atoms join in molecules, molecules and atoms make planets and stars, and stars in turn are the "atoms" that comprise galaxies. As though that surfeit of order were not enough to satisfy bountiful nature, galaxies themselves cluster by the dozens in round beehive swarms, called Abel clusters after their discoverer, each up to 10 million light years across. And these aggregate into superclusters—clusters of clusters. Our own Milky Way is only one galaxy in a local Abel cluster which is linked with about sixty other clusters of galaxies into a "local supercluster" more than 100 million light years in diameter. Other superclusters are filamentary and as much as a billion light years long—the largest organizations seen in the universe. At what level the hierarchies terminate, if at all, remains unknown.

Cosmologists believe that all this supra-atomic organization arose from the fine-structured undergrowth of the nearly featureless early universe. If the Big Bang had been perfectly uniform, the texture of the universe would have been boring and perhaps without recognizable organization called life. But physicists suspect that the seeds of emerging order were planted at the first instants of time, chance fluctuations in a fluid universe of colliding particles that later developed into stupendous organization. The undisputed *order* of galactic clustering is really an outgrowth of minor *chaotic* fluctuations amid primordial uniformity. Some of the random, quantum fluctuations may have occurred within the first 10^{-43} seconds, the Planck era of quantum gravity when all the forces of nature acted as one. Other chaotic excitation may have occurred later, but still within the first second of time, as the fundamental forces "froze out" or differentiated. An assault by cosmologists is now under way to trace the wellsprings of the hierarchical order of galaxies, a pattern that may be the signature of genesis writ large across the skies.

At first, cosmologists believed that galaxies congealed by gravity out of the firmament of primordial gas and then aggregated into clusters and superclusters. There are still those who believe this "bottom-up" cosmology, but a growing consensus imagines that the very largest scale clusterings that span hundreds of millions of light years were primary. According to this "top-down" model, galaxies then formed within the large-scale clumping that emerged from fluctuations at the beginning of time. The proponents of top-down say that insufficient cosmic time has passed to spawn large-scale structure by gravitational interaction. Perhaps the truth lies somewhere in between bottom-up and top-down, and both scales of organization occurred together or nearly so. One thing is certain, cosmologists are still grasping for an explanation of clustering hierarchies. There is yet no well-established theory.

One problem in galactic cosmology is the dearth and difficulty of observations. Less than 1 percent of the volume of space within a few billion light years of the Milky Way has been surveyed for the distribution of galaxies, even though telescopes with electronic sensors make observations far more rapidly than earlier techniques. Controversies have arisen many times over whether a pattern of large-scale clustering was a statistical fluke of sparse observations or was instead a real phenomenon. The absence of galaxies in certain regions of space—voids—have also turned up. Recent observations by investigators at the Center for Astrophysics in Cambridge, Massachusetts, reveal huge empty cosmic "bubbles" on the order of 100 million light years in diameter. Galaxies cluster on the surfaces of these cosmic voids, giving the impression of a vat of soap bubbles or a fluffy Swiss cheese.

As in many branches of science today, high-speed computers have been brought to bear on the distribution of galaxies. Cosmologists "play God" and simulate the unfolding order of the universe within the "minds" of machines. If the texture of the universe derived in a numerical simulation of interacting particles resembles the pattern of astronomical observations, then cosmologists gain confidence in the theory embodied in the equations and assumed initial conditions of their make-believe universe.

The cosmologists' models of the expanding cosmos focus on the clumping of matter on many scales of length that must have been well under way during the "matter era"—several hundred thousand years after the beginning when complete atoms formed and radiation lost its dominance. Some of the earliest clustering models were put forth by cosmologist P.J.E. Peebles and his associates at Princeton in the late 1960s and early 1970s. They proposed a bottom-up "isothermal" universe, one that had uniform temperature throughout as it expanded, and were able to evolve "protogalaxies" with 10^5 to 10^6 times the mass of the Sun. Out of these grew larger galaxies and then clusters of galaxies, though not the large-scale superclusters.

Separate studies by Yakov Zel'dovich in Moscow and Edward Harrison at the University of Massachusetts considered a top-down clustering model with only large-scale fluctuations. These so-called adiabatic models required that the density of matter ratioed to the density of radiation remain constant in each expanding region. The interaction of radiation and particles in such models prevents the initial appearance of small-scale clustering. Supercluster-sized regions form first and then subsequently spawn smaller zones of organization. Zel'dovich, for one, concluded that the early universe gave rise to huge pancake-shaped regions of gas and radiation. Where "pancakes" intersected, filaments of galaxies—superclusters—would have formed.

Various technical problems arose with these early top-down models that were inconsistent with the observation of a uniform background of cosmic radiation. Cosmologists realized that their simulations were failing to take into account the greatest bulk of the universe—the invisible dark matter. So began an era of simulations in which dark matter was incorporated, first by assuming it to be in the form of the fleeting neutrinos that had largely stopped interacting with protons and neutrons (generically called *baryons)* after the first 0.01 second of time. The neutrinos nonetheless were still so energetic in the early universe that, as shown by simulations, they erased small-scale structures but permitted larger ones.

While partially solving some of the technical problems of top-down models, the neutrinos—assumed for the sake of the models to have tiny masses, unlike massless photons of light—created other difficulties that may be impossible to overcome. Cosmologists Marc Davis at the University of California at Berkeley and Simon White of the University of Arizona have concluded from their simulations that neutrinos do not allow galaxies to form until much later than the ages of observed galaxies seem to indicate. At the same time, the neutrinos would introduce more larger-scale clustering than has been observed.

A new trend in cosmic modeling rejects the omnipresent neutrinos as candidates for the dark matter, assumes that they have negligible mass, and thus dispenses with their difficulties. Instead, the dark gravitating matter is assumed to be comprised of much more massive particles that are postulated in theories of "supersymmetry." Whether the particles are axions, gravitinos, photinos, or something else, they are characterized as slow moving or "cold," and not streaming freely through space in the manner of "hot," bulletlike neutrinos. They interact only gravitationally with normal matter. The new scenarios of cosmic evolution are called cold particle models and give bottom-up kinds of results—galaxies forming first, larger structures later. So far, neutrino models have not been able to give rise to the observed distribution of clustering and voids, but cold particle models can be made to work. Apparently both modeling and observations are in need of further development.

Cosmology and the behavior of particles in the microcosm are thus linked intimately not only in the first instants of the universe, but in the unfolding of galactic hierarchies over aeons. And perhaps cosmologists haven't considered all the possible contributors to large-scale cosmic order. Recently some investigators have suggested that filaments of galaxies may cluster near massive defects in the fabric of space-time that were present soon after inflation—very long so-called cosmic strings, that are one-dimensional "cracks" in space-time, as it were, related to other defects in space-time: monstrous *magnetic monopole* particles and *domain walls* that may border reaches of the universe far beyond our ken. Or perhaps the behavior of cold dark matter is much stranger than we can imagine, and new theories will be required for a comprehensive understanding of the heavenly tapestry.

Within the tangled spider web of galaxies lie the "atoms" of space—the stars themselves. A star is born when a huge cloud of gas and dust separates from the interstellar continuum by gravitational instability, contracting and drawing in vast quantities of fuel to stoke fusion fires that ignite when sufficient density and temperature have been reached. The stars are the inevitable constituents of the galactic whirlpools that spawn them. And as the starry "atoms" are born, so are planets accreted around them through complex processes that may well be the rule rather than the exception. The existence of copious extra-solar planets, long a great question mark in astronomy, grows ever more certain as evidence pours in from observatories on mountaintops and in space. Astrophysicists are forging the last link in the chain of being from the cosmos, through galaxies, stars, and planets. The sites for life as we know it—and as we don't know it—merely await the quickening pulse of matter and energy that are now far away from thermal equilibrium. As the universe slides into life, a watcher with a longevity of aeons would be struck by the continuity of universal evolution. The process of *quickening*, with the dual attribute of coming to life and accelerating, would be obvious and beautiful.

Life

6
THE NEW GENESIS

Is it possible that our race may be an accident, in a meaningless universe, living its brief life uncared for, on this dark, cooling star: but so—and all the more—what marvelous creatures we are! What fairy story, what tale from the Arabian Nights of the Jinns, is a hundreth part as wonderful as this story of simians! It is so much more heartening too, than the tales we invent. A universe capable of giving birth to so many accidents is—blind or not—a good world to live in, a promising universe.

—CLARENCE DAY, *This Simian World*, 1920

Given so much time, the "impossible" becomes possible, the possible probable, and the probable virtually certain. One has only to wait: time itself performs the miracles.

—GEORGE WALD, 1954

The emergence of life from the dust of the universe must be counted a great wonder, perhaps equal in glory only to the birth of the entire cosmos. Feeling awe about the origin of life admittedly may be a quirk of our limited perspective as brash upstarts—the young sprouts of the universe. Should older, wiser beings than ourselves exist, they may well take life's blossoming for granted as the inevitable transformation of matter. After much more probing and poking at the substance of terrestrial life, we may arrive at that viewpoint too, but in our present gasping state of ignorance life's beginning has an aura of the miraculous. Perhaps this may be fortunate, for the childhood of a species or of an individual should be full of seeming miracles and new mountains to climb, not jaded with complacency.

Determing the ease or difficulty with which life began on Earth is critical to knowing whether we inhabit a universe that is otherwise mute—an incomprehensible asymmetry, considering the countless quadrillions of stars and planets—or whether the cosmos is veritably teeming with organization. If life is the legitimate offspring of matter, given time and the right circumstances, we are indeed not alone in pondering the world and may someday encounter fellow questers. What process breathed fire into senseless matter is a question that science cannot yet answer, but the scientific pursuit of the origin of life has uncovered tantalizing clues and unleashed new controversy.

The culminating celestial fact of which there is little doubt is that we are all star stuff. All the atoms of our bodies were cooked in primordial nuclear fires at the birth of time or blasted from the innards of dying suns. We come most literally from the stars, but where was the incubator that hatched life from atomic simplicity? Not only the *how* but the *where* of life's beginning is the focus of the scientific quest in laboratory and field. The prevailing scientific model is that life began in an ocean of chemical broth cooked up on the primitive Earth more than 3.5 billion years ago. It is a plausible and romantic story of life sparked by lightning, heat, and radiation, one whose earthly setting at least may be correct. But there are stirrings of dissent among those who say terrestrial life may have begun elsewhere, perhaps on worlds far removed from the Sun and its brood of planets. Unlikely as the minority view seems, it deserves a modicum of scrutiny.

The overwhelming biological, geochemical, and fossil evidence reveals that the course of life has followed the path of evolution since its inception, sometimes slowly and at other times in apparently great spurts. Scientists still argue heatedly about the mechanisms of biological evolution, but that life evolved to its present state once it acquired genetic apparatus is as close to scientific fact as anything can be. If one dogmatically asserts that on Saturday evening, October 22, in 4004 B.C. or thereabouts, the universe came into being by Executive fiat and that all life appeared fully formed that same week, so be it. But that stubborn *belief* must never be confused with science. Science is a *method* for ascertaining facts about existence. Those facts developed by science are merely the reports of observations and experiments, together with logical inferences, all of which are always subject to revision. The record fossilized in the rocks and in the genes of every organism on the planet doesn't lie. There is not a shred of scientific evidence that denies that life evolved, yet preposterous clouds of confused dogmatic belief threaten here and there to eclipse one of the most beautiful stories science has ever revealed.

Evolution of primitive bacteria in the direction of multicellular life is remarkable enough, but the origin of the first viable replicating or metabolizing cell seems much more incredible and remains a deep scientific mystery. The great paradox of modern science is that it knows more about the first second of the Big Bang than about the first billion years of presumed molecular evolution on Earth that led to the first cells. Science is even far from certain that life as we know it is the only type of "life" that may grace the cosmos. Many twists and forks in chemical evolution on the road to life might have led to substantially different biochemistries, even within the domain of organic chemicals—molecular structures with carbon atoms as central features.

Besides the carbon-based biochemistry with which we are familiar, other unusual chemistries may also lead to replicating and information processing structures. Even more difficult to imagine, perhaps other "life" inhabits niches totally alien to conventional understanding—the wastelands of frozen planets, the hot surfaces of dense stars, or interstellar space itself. Science is now in the uncharacteristic position of studying but one kind of life and has nothing with which to compare it. After probing the mystery of terrestrial molecular complexity and how it managed to gain control of a planet, we'll pause and imagine that the universe might be "biotropic." It may delight in inventing endless multifarious organizations along pathways so alien that they may appear not as life but as sheer wizardry.

Genesis One

By radioactive dating of meteorites and other evidence, we have determined the age of the solar system to be about 4.6 billion years. The Sun and its retinue of planets condensed by gravitational attraction from a gigantic cloud of gas and dust, similar to those observed by astronomers even today in the process of forming new stars and, very likely, planets in attendance. The oldest rocks on the surface of the Earth, found in Greenland, formed 3.9 billion years ago but experienced too much heating for any possibly fossilized cellular life forms in them to be apparent. Yet analysis of the nuclear forms of carbon in those rocks is consistent with the idea that they are residues of bacteria.* But some rocks 3.5 billion years old definitely have the fossil remains of single-celled organisms. The startling conclusion: life appeared on Earth about a billion years after the planet solidified, a small fraction of its age, an even smaller fraction of the age of the universe. How much less than that billion years it took to get to the very first cell may never be known, but it could have taken a surprisingly short time—several hundred million years or even far less.

The nature of the earliest creatures 3.5 or more billion years ago, whether or not they possessed genetic apparatus, is not certain. But the relatively short time for the first cell to appear

---

*See *Microcosmos* by Lynn Margulis, and Dorian Sagan, p. 72.

may suggest that life starts readily on a suitable planet. If genesis were an extremely chancy affair, the first cells might have taken much longer to form. Whether life took 100 million years or a billion years to begin, those times are as flashes compared to the pessimistic estimates of time needed "accidentally" to jostle molecular fragments into life—billions of times the age of the universe. (A devil's advocate would suggest that Earth is simply one of the rare planets in a virtually infinite universe where the process happened quickly.) Contrariwise, the succeeding approximately 3 billion years, through which the fossil record shows life remaining dominantly unicellular, indicates that complex multicellular life may require a more lengthy synthesis as it emerges on other worlds.

Only 570 million years ago, at the beginning of the so-called Cambrian period, life began an evolutionary explosion that led to great diversity of plants and animals in water, on land, in air, and just now in space. As recently as the 1950s scientists in fact assumed that all life had begun about that time, overlooking that fossilized remains of earlier gelatinous bodies or microbial cells would be more difficult to find. The course of evolution beginning 570 million years ago is a fascinating and controversial subject in itself, but the preceding billions of years of microscopic evolution set the stage. It was a gestation period during which tiny invisible beings, multiplying wildly and interacting parasitically and symbiotically, transformed the atmosphere of an entire planet. The biochemical processes invented by early life not only persist in microbes today, but are the foundations of the most complex life-forms built of quadrillions of cells.

Science tries to be immune to infectious mythology, but it doesn't always succeed in warding off the disease. So difficult is the scientific quest for the origin of life that it has understandably lapsed into mythology now and then—myths not in the sense of concocted fables, but the creeping confusion of scientific plausibility with established facts. Ask most scientifically minded nonexperts today and they will probably tell you that the chemically primitive units necessary for life readily formed and gave rise to a "primordial soup" on the young Earth. They will then tell you that the weight of modern scientific opinion is that life started in the oceans, a chance or inevitable outgrowth of the prebiotic soup of organic molecules jostling and merging. This isn't far from what most scientists believe—a nice theory, one that *may* be true, but also one that is far from being demonstrated to be correct. Yet good scientific myths about tough puzzles have their advantages. They are positions against which to compare other evidence and alternative theories. So with a little background music for life's entrance onto the stage, let's plunge into the primordial soup.

Life as we know it consists of cells or a cell whose main characteristics are an ability to metabolize—exchange chemicals with their surroundings to get energy and material for their workings—and an ability to pass on information to succeeding generations of similar cells. The handing down of the plans of life is the responsibility of nucleic acids within cells, DNA, which is a long, ladderlike, double-stranded and twisted molecule—the repository of the genetic code, and RNA, which is a single-stranded molecule that helps carry out DNA's plans. (Viruses and viroids are a bit outside this realm, having no ability to metabolize, and are mostly pure genetic material combined with some protein structure to help them invade metabolizing cells. They are thought to have evolved after the advent of cells, and are perhaps the errant parasitic fragments of cellular nucleic acids.) It is by no means certain which characteristic of life originated first: metabolism or the genetic code. We now have the first glimmerings of how these functions could have developed, though it wasn't long ago that science could hardly imagine how even chemical precursors of genetic material and proteins could form from simpler inorganic matter.

The Russian biochemist A. I. Oparin and British biologist and geneticist J. B. S. Haldane independently suggested in the 1920s that the primitive Earth did not have an oxygenous atmosphere, but rather had what is called a *reducing** one made of gases such as hydrogen, methane, ammonia, and carbon dioxide. Moreover, they postulated that this atmosphere might lead to the formation of primitive organic molecules. It wasn't until 1953 that their suggestions were subjected to test, in the seminal Miller-Urey experiment. A graduate student, Stanley Miller, acting on the advice of his professor, Harold Urey, at the University of Chicago, set up a glass chamber into which methane, ammonia, hydrogen, and water were introduced. He allowed a high-voltage electric spark to disturb the heated mixture, such as lightning might do on the young Earth. Vapors from the chamber cooled and then recirculated. Within a week Miller was able to find the building blocks of proteins in the chamber—amino acids plus other organic molecules. This was the prototype of the mythic primordial soup, the first of many to come.

In the same year Francis Crick and James Watson finally unraveled the mystery of the replicating nucleic acids by finding that DNA's structure consisted of an interlocked double helix—looking for all the world like a spiral staircase—whose two strands could be separated. With the aid of the cell's molecular machinery, the disentangled fibers could reform themselves, using molecular spare parts, resulting in two identical copies of the original genetic molecule. By the early 1960s biochemists discovered that by a simple coding sequence of four distinct so-called nucleotide molecules comprising the DNA

---

*A *reduction* chemical process is one in which an electron is *added* to an atom or an ion.

double helix, complex information could be stored and transferred, analogous to the way data resides in a digital computer.

Each series of three nucleotides in the DNA forms a code for a single amino acid molecule. Nature inscrutably selected a remarkable set of twenty amino acids out of many thousands of possible related molecular forms (including some fifty bypassed amino acids) to be the units that DNA specifies to be sequentially strung together to create a protein. A length of DNA nucleotides that codes for the hundreds or thousands of amino acid units that make up a single protein is known as a gene. The amino acid stringing is done under the control of forms of RNA that transfer DNA's message to other parts of a cell's molecular machinery—units called ribosomes.

While the Miller-Urey and Watson-Crick developments were extraordinary, no living creature has of course leaped or crawled out of a test tube primed with simple organic molecules from primordial soup. It has not been possible to concoct DNA or RNA from the primitive molecules coming from Miller-Urey flasks, because those stewing pots have never brewed any of the critical complete nucleotide subunits. Nor have investigators found any amino acids strung together to form conventional proteins.

Biochemists attempting to illuminate how life might have formed are faced with a chicken and egg problem at many levels. In order to make proteins, a cell needs amino acids and the key ingredients of cells, enzymes, which act as catalysts in transferring DNA's blueprint for the protein to be made. A catalyst is a molecule that acts like a key to facilitate a chemical reaction, but the catalyst itself is unaltered by the reaction. Without the key, no reaction. The problem is that enzymes themselves are proteins—how does one enzyme get made before another enzyme required to make it exists? This is one of the most fundamental but not the only problem in discovering how life began. Enzymes are also needed to permit DNA and RNA to replicate. Biochemists have not been able to make enzymes by accidentally agglomerating the right subunits—the probability of that happening even for one kind of enzyme is vanishingly small, given even an "ocean" laboratory flask much larger than the combined seas of many worlds and time greater than the age of the universe.

Some take biochemistry's failure so far to unravel the mystery of the origin to be a sign that supernatural intervention, or freak chance equivalent to such, must be invoked. But this is not the opinion of most biochemists, who do not really hope to create a cell from primitive molecules, but who, in the words of biochemist Richard E. Dickerson, have the "broad goal to arrive at an intellectually satisfying account of how living forms could have emerged step by step from inanimate matter on primitive Earth." Biochemist Leslie Orgel states that "we have clues that the sorts of chemistry that are needed for the origin of organized systems can occur—so that it is clear that [the origin of] life is perhaps not a miracle, but nevertheless we don't understand much about it." Orgel graphically illuminates the source of our ignorance: "Imagine a professor, very long lived and a master of getting research grants, and working in his lab for 100 years with 100 graduate students day and night, each of whom have 10 liters of various chemistry cooking continuously. That would only amount to 100,000 liter-years of effort. The primitive ocean of Earth with a billion years at its disposal had 10^{29} liter-years."

Perhaps in that early ocean of Earth an original "naked" self-replicating molecule did arise which began to dominate the chemical stew. Unshielded from the surrounding medium, it is hard to imagine how such a chemical system could preserve itself very long. So some biochemists have experimented with creating primitive "cells," mere pockets of chemical broth surrounded by a fatty coating. Whatever chemical innards these sacs came to nurture would not be in danger of being diluted to oblivion. To see how they behave, scientists have infused biochemicals into some of these microspheres, which have the same size range as real cells. The fake cells wondrously acquire lifelike properties such as stability by virtue of their primitive "metabolism" and an ability to divide and produce "offspring." It is a very long leap to the first real cell, but each new experiment casts light on what may have happened long ago.

The New Genesis

The genesis mythology of the last three decades requires that a "naked gene" or self-replicating molecule was given difficult birth in the primordial soup. As this unclothed "gene" multiplied, it began a process of molecular evolution that led to cellular life. For all our ignorance, this may represent what actually happened, but even the first chapter of the myth has not remotely been enacted in the laboratory. First, the genetic nucleotide bases are synthesized only with great difficulty and in low concentration in Miller-Urey type experiments. Second, given an elaborately contrived "soup" of complete nucleotides, they do not link together without the molecular assistants called enzymes, and then dance happily as a naked gene. The naked gene assertion has not been completely without experimental support, but those experiments violate the requirement that there be no improbable molecular helpers in the soup.

A rebellious faction has arisen against the naked gene or "replication first" paradigm by those seeking a more plausible genesis. Freeman Dyson, for one, has examined the origin of life question from the unbiased vantage point of a physicist watching biochemists and biologists doing battle. He suggests a way out of the chicken and egg dilemma of genesis by proposing that life had a dual origin. According to his hypothesis, a purely metabolic creature first arose, one incapable of the precise replication we see today in the fissioning of DNA-governed cells. Nevertheless, chemical equilibrium established within an oceanic community of fatty sacs enveloping primitive protein molecules may have been adequate to sustain a kind of natural selection of the fittest of these microscopic blobs. Dyson maintains that exact replication would not have been essential for Darwinian-like evolution of the microspheres. Since primitive proteinlike molecules, through the heating of amino acid mixtures, are more readily synthesized than the nucleotide building blocks of DNA or RNA, the natural genesis of proteinous metabolizing forms is more credible than the advent of naked genes.

The second phase of the origin of life theoretically occurred with the chance incorporation of nucleotides into the proteinous sacs—the parasitic arrival of gene building blocks. No doubt the more confined volume of the primitive protein microspheres was conducive to all manner of enzymatic activity that jostled the nucleotide building blocks into the replicative structures they eventually became. Dyson and others have suggested that RNA first appeared and later helped to engender its future master, DNA—like the child giving rise to the parent!

Manfred Eigen in Germany has already demonstrated a key aspect of the dual origin theory. In his experimental broth of nucleotides assisted by the presence of a single enzyme obtained from bacteria, Eigen has created a population of growing chains of nucleotides—artificial RNA. Moreover, these RNA molecules literally evolve in the experimental chamber through replication, mutation, and competition. Admittedly, Eigen has "cheated" by introducing a critical bacterial enzyme, but this is nonetheless a significant demonstration of primitive molecular evolution.

Leslie Orgel at the Salk Institute in the United States has performed supporting experiments to Eigen's, demonstrating that under certain contrived conditions nucleotide building blocks if given an RNA "template" will copy that RNA *without* enzyme assistance. But to imagine parasitic genesis of

RNA inside the primitive metabolizing cells, we'll have to make do without either Eigen's bacterial enzyme or Orgel's RNA template—a feat no one has yet accomplished. Yet who can say that the next twenty years of biochemical experiments will not demonstrate RNA genesis without trickery?

In the quest to understand genesis, the argument between the naked gene theorists and the metabolism firsters goes on. The paradigm of the stripped-down gene in the organic soup won't easily disappear. But Dyson's grand scheme of dual origin synthesizes the two viewpoints and recapitulates some of Oparin's thinking earlier in the century, before the structure of genes was known. Dyson summarized his views connecting genes to homeostatic metabolism and concluded, "In my version, the history of life is counterpoint music, a two-part invention with two voices, the voice of the replicators attempting to impose their selfish purposes on the whole network, and the voice of homeostasis tending to maximize diversity of structure and the flexibility of function."

Other streams of thought refresh the search for life's beginning. Graham Cairns-Smith in Scotland has propounded his own intriguing dual origin of life theory, which he describes as "genetic takeover." But his first stage of genesis involves crystalline clays which he imaginatively suggests already possess primitive self-replication properties. The essence of crystalline lattices is, of course, replicative, as layer upon layer of atoms automatically fall into place guided by the electromagnetic geometry of atomic affinities. He says that current life is too "high-tech"—meaning too complex for naked genes to form in primordial seas.

Imagine "low-tech" crystalline patterns acting like genes to preserve information. When water flows through porous crystalline clays, crystal fragments—genes—may break off and travel downstream to begin new sites of crystal life. We can conceive of an evolutionary process whereby "metabolizing" crystal genes proliferate and adapt to the environment. Cairns-

Smith postulates that at some stage of this stony evolution organic molecules, that naturally affix themselves to the gene-like crystals, took over the job of molecular evolution, using the crystals as templates. All life might derive from that first molding of carbonaceous matter through the geometries of silicon architecture. Genetic takeover between crystals and the organic molecular precursors of present life is the essence of this new story of genesis. Cairns-Smith suggests that remnants of the crystal stage of life might still inhabit some protected ecological niche and we might seek this common ancestor of the quickening world.

Theories of dual origin and crystal genesis expand the horizons of possible explanations for life's seemingly miraculous advent. Perhaps other pathways leading back to primeval simplicity still lie buried in the complex overgrowth of advanced life. Whenever we see any kind of order emerging from chaos in the chemical world, our life detection alarms should sound. There are chemical reactions that spontaneously evolve to organized macroscopic states that oscillate in time and space. One, called the Belousov-Zhabotinsky reaction, generates extraordinary, unexpected, pulsating geometric patterns in thin films. Chemists have suggested that this reaction and its relatives may be models for cyclic chemical patterns that may have been the first instances of self-replication. With all these possibilities as prologue, before anyone insists that genesis must have been a "miracle," one forever unexplainable in normal scientific terms, their evidence ought to be so solid that finding falsehood in that evidence would be a greater miracle.

Microcosmic Evolution

What biochemists lack in understanding of processes leading to the first cells is made up for by an increasing knowledge of the steps of cellular development that led to the evolutionary explosion of multicelled organisms later on. The first cells were prokaryotes, cells that did not have a central nucleus to contain their DNA. The fossil record shows that early prokaryotes are closely related to contemporary bacteria and blue-green algae, also called cyanobacteria. The earliest prokaryotes must have had an anaerobic metabolism, that is, they got their energy without relying on an oxygen atmosphere, using a process akin to contemporary fermentation. They probably didn't require sunlight. Some have speculated that early bacterial life may have emerged near undersea thermal vents that spew forth hot water and sulfurous gases. In recent years bacterial colonies and larger organisms have been found near these cracks in the Earth's crust, an ecology that seems totally independent of sunlight's daily caress.

Some of the early bacteria then developed an ability to use sunlight to store energy chemically—photosynthesis. Thus, the second prokaryotic stage was "anaerobic photosynthetic." Then some cells began to tolerate the new poison which they had begun to release to the atmosphere—oxygen. Oxygen was a deadly toxin to the anaerobes, much as it is today. The new cyanobacteria using aerobic photosynthesis created the massive amount of oxygen in the atmosphere. The oxygen in every breath we take is a result of this bacterial processing. Oxygen permitted an ozone layer to form in the upper atmosphere, thus shutting out ultraviolet radiation which would be destructive to surface life. Ozone is a molecular form of oxygen that combines three atoms instead of the usual two. Prior to life's fortuitous creation of the stratospheric ozone layer, unfiltered ultraviolet radiation perhaps contributed energy that helped form the prebiotic soup—a marvelous synchronization of events, to say the least.

The period of exclusively prokaryotic cells lasted to about 1.4 billion years ago, when the first eukaryotes appeared—cells with distinct nuclei containing their genetic material. It is not always understood how much smaller prokaryotic bacterial cells are than eukaryotes. If a bacterium were enlarged to the size of a person, a typical single cell eukaryotic organism would be as large as the Washington Monument—about 170 meters. This dimensional relationship adds credibility to the very strong evidence that links the advent of eukaryotes to a parasitic symbiosis among prokaryotes. The various organelles of eukaryotic cells— the nucleus, energy-producing mitochondria, chloroplasts responsible for photosynthesis—may well have started as prokaryotes that took up residence within other prokaryotes. The significance of the eukaryotes is that they were the first organisms able to reproduce sexually. Prokaryotes reproduce asexually by simple fissioning. Sexual reproduction, with its ability to produce more readily slightly altered forms in each generation, no doubt greatly speeded up evolution. Significantly, the greatest evolutionary expansion occurred shortly after eukaryotes appeared. It has been said that "the price of sex was death," for in their fissioning reproduction prokaryotes eternally produce identical copies of themselves (except for chance mutations) and are the closest thing to immortal life. The deaths of eukaryotes and higher forms continuously make room for new forms of life.

Panspermia

As far back as Darwin, the "warm little pond" notion of life's origin has held sway. In 1871 Darwin wrote: ". . . if (and oh! what a big if!) we could conceive in some warm little pond, with all sorts of ammonia and phosphoric salts, light, heat, electricity, etc., present, that a protein compound was formed ready to undergo more complex changes." But what if that little pond was not on Earth? Could terrestrial life have originated elsewhere, and does it matter?

Since the Swedish chemist Svante Arrhenius proposed the theory of panspermia in the first decade of this century, these questions have refused to go away. Arrhenius contended that spores of life could rise in the atmosphere of another planet, either in this solar system or around another star. Electrostatic forces might eject such spores from the planet's atmosphere and the spores could waft across interplanetary or interstellar space to start life on other worlds. Perhaps life on Earth began this way.

Panspermia would not remove the thorny problem of the biochemical origin of life, just change the site of genesis to another world at an earlier time. The origin of life is, after all, more like history than physics—exact calculation of molecular events very remote in time may be impossible, yet the sequence and location of occurrences is of great interest. Carl Sagan and others have cast doubt on the panspermia hypothesis by calculating that unshielded microbes could never survive the ultraviolet, X-ray, solar proton, and cosmic ray irradiation of long space journeys in vacuum and minus 196° C cold. Experimental work by J. Mayo Greenberg at the University of Leiden seems to reopen the case, yet the difficulties for natural panspermia are still formidable. A vast quantity of spores would have to emerge from another star system to have a good chance of reaching Earth, yet in theory just one spore would be sufficient.

Francis Crick and Leslie Orgel proposed a more radical theory—deliberate or directed panspermia on the part of intelligent aliens. This theory is much more difficult to dispute, since it is conceptually possible that an advanced space-faring race would have had the technology of interstellar transport to seed primitive Earth with bacteria. Whether there exists any reasonable motivation for this directed panspermia (territorial passions, egocentric urges, the search for biological immortality?) is an open question. Physicist Thomas Gold once suggested that an early landing expedition by aliens may have inadvertently contaminated Earth. A "space garbage" origin seems quite opposite to the inspiring romantic vision of life emerging from the sea. Crick and Orgel propose a number of reasons to believe that directed panspermia did occur, though are careful to add that they suggest the idea only as a theory that is *consistent* with scientific knowledge.

As evidence for directed panspermia, Crick and Orgel cite the universality of the genetic code on Earth. If the Earth was replete with warm little ponds, wouldn't other forms of the code have arisen and persisted? Directed panspermia easily dismisses this problem—the common ancestor of life was aboard an alien spacecraft. They point to other anomalous aspects of biochemistry; for example: the element molybdenum is an essential ingredient, yet is a minute trace constituent in Earth's crust. Perhaps the strongest reason to believe intentional seeding occurred is the rapid start that life achieved after Earth formed. A more ancient populated world could have recognized an incipient hospitable planet and acted immediately to spread their heritage.

Life may not have arrived from space as fully formed microscopic organisms, but the chemical precursors of life might have come from interstellar space. Radio astronomers are today faced with a bewildering host of microwave signals coming from dozens of organic (carbon structured) molecules in space, including formaldehyde, hydrogen cyanide, and precursors of amino acids. Some think that these organics might have been deposited on Earth through collisions with comets, icy wanderers between the stars. There is confirmed evidence that organic molecules as complex as amino acids have arrived on Earth from meteorites. The Murchison meteorite that landed in Australia in 1969 was found to have equal proportions of geometrically "left handed" and "right handed" shaped amino acids—a clear indication that processes in space formed them and that they are not simply contaminants from terrestrial life. Earth life universally incorporates only left-handed amino acids—a pervasive bias that is a strong case for life having had a common biomolecular ancestry.

Astrophysicist Fred Hoyle and his colleague Chandra Wickramasinghe have startled—and many would say appalled—the astronomical community with their theory that life is co-eternal with the universe and proposing evidence to support it. Hoyle is one of the originators of the now widely disbelieved steady state cosmology that holds that the expanding universe is infinitely old, that matter is formed continuously from vacuum, giving the universe roughly the same appearance at all epochs. Hoyle and Wickramasinghe begin with the premise that it is impossible to imagine the chance aggregation of nucleic acids and associated enzymes required to build proteins. The mathematical probability, as they compute it, is far too low. Hence, life in the universe is ordained by a "pre-existing intelligence" in their view. The consequence of this notion is that viral, bacterial, and even insect life is conceived to populate open space. This life wafts down on planets and starts an evolutionary adventure that continues to be excited by interstellar viral influence. The authors of a series of technical papers and books on the subject point to geographic patterns of Earthly disease epidemics as evidence for their theory. Their alleged strongest evidence, still much disputed by others, is the claim that microfossils of bacteria are contained in several well-known meteorites. They also cite evidence even more strongly rejected by most astronomers: ultraviolet spectral emission of interstellar compounds of cellulose, almost literally "vegetables in space." In his book *The Intelligent Universe*, Hoyle seems to have taken flight into still more metaphysical realms as he speculates on the hierarchies of incomprehensible intelligences in the cosmos that impose life on virgin planets.

Before following life further up the chain of being, it is worth noting that Charles Darwin at one time considered the origin of life almost beyond the pale of science. He wrote in a letter in 1863, "It is mere rubbish thinking at present of the origin of life; one might as well think of the origin of matter."

He would be surprised at the enormous strides science has made in understanding both of those questions. Were we to live 100 more years, it seems certain that, looking back on today's penetrating but still primitive understanding of genesis, we would experience the same incredulity.

The majestic spectacle of the universe quickening—arriving at self-knowledge, first through molecular and then cellular evolution—is infinitely inspiring to those who study it. Scientists are profoundly in awe of what they have already seen in the natural record, ironically making the stubborn, groundless idea of the instantaneous creation of life all the more blasphemous to this subtle, beautiful, and mysterious history. Someday we may know with certainty whether life is the inevitable child of the physical universe or is a chance event almost unique to our small quarter. The vote now in the halls of science is dominantly on the side of inevitability.

7

GAIA

If a germ cannot be presumed aware of the living state of the body it dwells in, how can man's somewhat similarly circumscribed view afford him much more comprehension of the total aliveness of his planet today. . . .

The physical essence of Earth life may be termed a spherical biofilm rotating in gravitational, electromagnetic, and nuclear fields—a sort of gyrating bubble of evolving potency, a cosmic node of ferment.

—GUY MURCHIE, *The Seven Mysteries of Life*, 1978

The clues to Gaia's existence are as transient as our sand-castle. If her partners in life were not there, continually repairing and recreating, as children build fresh castles on the beach, all Gaia's traces would soon vanish. . . .

The Gaia hypothesis is for those who like to walk or simply stand and stare, to wonder about the Earth and the life it bears, and to speculate about the consequences of our own presence here.

—JAMES LOVELOCK, *Gaia: A New Look at Life on Earth*, 1979

A singer romantically intones on the local country and western music station, "You're the reason God made Oklahoma . . ." —an unwitting electromagnetic message to the universe that life and love has evolved on a once virgin world in the backwaters of the Milky Way. If alien recipients could interpret the radio message, they might conclude mistakenly that Earthlings once had an intense interest in predestiny, the philosophy often ascribed to the anthropic principle!* The world is indeed alive and the evidence surrounds us, radio broadcasting being but one of numerous tendrils of life suffusing the planet

*See Chapter 4.

and beyond. But is it "alive" only in the metaphoric sense of *bearing* life, or does the planet actually live as a global super-organism, comprising the totality of all that lives and even that which appears not to?

Many ancient peoples, more tuned to the myriad cycles of life than modern city dwellers, never doubted that all nature was part of an Earth Spirit—that the Earth was in fact alive. To the ancient Greeks the spirit was the earth goddess they called Gaia. The more that modern science learns about life's secret ways and the planet that biology has conquered, the more tenable becomes a holistic view of Earth—a rational version of Gaia. It grows increasingly difficult *not* to conclude that the atmosphere, the seas, the land, and all the biota they contain have acted symbiotically through aeons as a single living organic complex.

The biosphere consists of the continents, the atmosphere, and the seas—an incredibly thin layer of matter coating a stony world and its underlying molten iron-nickel core. The part of the atmosphere we breathe, the domain of weather called the troposphere, extends upward only 12 kilometers and contains more than 75 percent of all air. Comparatively, it is thinner than a sheet of paper covering a small melon. From the top of the troposphere to a depth of several kilometers below sea level, the soft biosphere of air, water, and rocky continents averages only one-third the density of water, whereas the whole Earth has a density more than five times that of water. This wondrous, pulsating biofilm has apparently stabilized the average global temperature near 22° C for hundreds of millions, perhaps billions of years, even as the Sun has grown brighter by some 30 percent since the origin of life.

The rebirth of awareness of Gaia occurred in the 1960s and 1970s, ironically as researchers planned expeditions to search for life on another world—Mars. British scientist James Lovelock, who had been a consultant to NASA on life-detection experiments, realized that we might be able to find evidence for life on Mars without even going there. He reasoned that the atmospheric composition on a life-bearing planet could be a telltale sign if life had distorted the mixture in a characteristic way. Examining the spectrum of reflected light from a planet like Mars would reveal much about its chemical constitution, and therefore perhaps whether it had life. This motivated Lovelock to consider some unusual features of Earth's atmosphere that were possibly connected with life. He soon realized that life must be controlling the atmosphere in ways never before suspected. These speculations led him to formulate the "Gaia" hypothesis, the name for the new conception of Earth suggested by his neighbor, novelist William Golding, at Lovelock's request. In his 1979 book *Gaia: A New Look at Life on Earth* Lovelock explained his revolutionary scientific idea that harkens back to ancient folk wisdom. In summary, "The entire range of living matter on Earth, from whales to viruses, and from oaks to algae, could be regarded as constituting a single living entity, capable of manipulating the Earth's atmosphere to suit its overall needs and endowed with faculties and powers far beyond those of its constituent parts."

The Gaia hypothesis violates the conventional scientific wisdom that life merely adapted to pre-existing terrestrial conditions, perhaps making a few modifications in the environment, such as giving rise to oxygen in the air, but being in essence merely a rider on Buckminster Fuller's "spaceship Earth." By contrast, the Gaia hypothesis suggests that life has played an active role to sustain itself by modifying and controlling the environment. Scientists involved in the study of ecology on a small scale have long been aware of the intricate interdependencies of life-forms, but the notion of strong global linkage and almost "purposeful" modification of the environment was difficult to accept and is a continuing controversy.

Since James Lovelock proposed that Gaia grows with us and through us, scientific investigations have borne out a host of interrelationships between life and the environment. Instead of two essentially independent evolutionary processes, that of the terrestrial environment and that of life, it is clear that the planet and life are co-evolving, constituting a unified evolutionary scheme. That this symbiosis occurs not only blurs the distinction between life and nonlife but may be a hopeful indication that similar activity might sustain life on other worlds. It is difficult to agree with the view of some skeptical scientists that the biosphere is simply a lucky concatenation of adaptations that at many turns could have spelled doom for all life. Life accidentally blundered its way through on one small world, while failing miserably perhaps on numerous now sterile planets. It seems instead that there may be a fundamental and perhaps universal process operating within the biosphere. Biologist Lynn Margulis has said in support of Gaia that it seems that Darwinian natural selection of organisms constitutes the fundamental feedback process that drives the superorganism. In other words, there is an underlying Gaian mechanism for what otherwise would have been regarded as pure mysticism or else blind chance. Lynn Margulis is not particularly happy with the term "superorganism," feeling that it may confuse people about the underlying Gaian mechanisms. But her misgivings, it seems, are based primarily on semantics.

Science has probably only scratched the surface in understanding Gaia, even though volumes have been written about what we do know. We must go back over 4.6 billion years to Gaia's birth when the planet began as a molten aggregate of rocky planetesimals congealed by gravity from the original solar nebula. It began to cool and the newborn Sun swept clouds of hydrogen and helium from the planet's clutches. As the heat of gravitational melting and radioactive decay caused many heavy elements like iron and nickel to settle toward the center, gaseous eruptions and belching volcanoes coughed up the primitive atmosphere of lighter elements and oceans of water. Those early conditions didn't allow oxygen to free itself from the bondage of chemical combination with other elements in the body of the Earth, yet today the atmosphere is a staggering 21 percent oxygen—the chemical hallmark of life's subsequent transforming action.

Before the origin of cellular life, the primitive biosphere may have had self-regulating chemical cycles that maintained clement temperatures, preventing the oceans from freezing or boiling away. Some scientists have speculated that quantities of ammonia, methane, and carbon dioxide gas in the early atmosphere acted as a complex thermal blanket to trap solar radiation. Earth's atmosphere at the outset may have had more than 100 times its present minute level of carbon dioxide or CO_2, a gas that exerts a powerful influence on global temperature. In what is known as the "greenhouse effect," CO_2, like other gases that have a similar property, traps the light energy of the Sun by preventing it from being re-radiated back into space as infrared (long wavelength) radiation. Through the greenhouse effect, the dense CO_2 atmosphere of Venus has made it a hellish world of high temperature and pressure, where lead could flow in molten streams. The much thinner Martian atmosphere, though predominantly CO_2, has made that planet much colder than Antarctica.

Gases from the atmosphere naturally dissolve in the oceans. Today, for example, the seas contain in dissolved form perhaps 50 times the amount of carbon dioxide that exists in the air. In the early atmosphere not only CO_2, but other greenhouse gases such as methane and ammonia, must have cycled between the sea and the air as temperature varied with the Sun's activity and the Earth's orbital motion. When excessive cooling occurred, more greenhouse gas could dissolve in the ocean. Other complex geochemical factors could make the gases escape to the atmosphere, leading to a warming adjustment. The stability of this planetary scale homeostasis—the analogue of the human body's chemical, pressure, and temperature regulating systems—may well have been very fragile prior to the advent of life. An excessive chemical or temperature excursion in one direction could have tilted conditions permanently to extremes inimical to the development of life. But once life started and spread like wildfire to fill every ecological niche, its robust Darwinian vitality allowed it to "invent" metabolic pathways to augment the primitive Gaian control systems that were then mainly cycles of trial and error.

The evolution of microbial life continues its forceful regulation of atmospheric and oceanic composition since the time of the first cells. About two billion years ago a startling toxic pollution episode occurred that sorely tested Gaia. Microbes started to emit volumes of oxygen, a toxic gas if ever there was one to unprepared life-forms. Microbes that couldn't tolerate the noxious gas learned to live underground, and the oxygen-loving (aerobic) bacteria inherited the oceans, the surface lands, and the air. Surviving this greatest pollution crisis of all time, Gaia transformed herself into an even more potent being with a free-floating energy reserve of oxygen.

Prior to the 1950s scientists thought that most of the Earth's oxygen was the product of the breakdown of water by ultraviolet light from the Sun. The idea was that a water molecule, consisting of two hydrogen atoms and one oxygen, split apart and the fleeting light hydrogen atoms leaked out into space, leaving behind oxygen. But how was this massive "sea" of oxygen to be maintained against its constant tendency to combine (combust) with virtually all the surface materials lying around? The atmosphere of 21 percent oxygen, 78 percent nitrogen, and the balance of trace gases is presently far removed from what is called an equilibrium condition, a state in which all chemicals have reacted and fallen to their lowest energy state. On an equilibrium Earth the atmosphere would be almost entirely CO_2 with not a trace of oxygen or nitrogen.

The seas would also dramatically increase in salinity. The biosphere is kept safely away from this deadly state by the combined metabolism of all biota. Living plants and algae, of course, give off prodigious quantities of oxygen. And their continuous death and decay also sends oxygen into the atmosphere and carbon (from carbon dioxide) to safe burial at the bottom of the sea. Without life, without decay, oxygen would gradually disappear from the air and life would have to return to its earlier anaerobic bacterial state before the great oxygen poisoning.

Oxygen sustains life but it also has brought the planet perilously close to the brink of conflagration. Experiments show that if oxygen concentration were to increase a few short percentage points and reach 25 percent, even damp vegetation would disappear in a whirlwind of fire ignited by chance lightning strikes. But anaerobic bacteria in marshes, seabeds, and in the guts of termites annually produce more than a billion tons of methane gas that combines with the oxygen and reduces its concentration. Were it not for the bacterial methane sink for oxygen, scientists have calculated that oxygen would increase a hazardous single percentage point in roughly 10,000 years. On the other hand, too much methane would be undesirable, because below 15 percent oxygen concentration no flame could be sustained; both forest fires and campfires alike would be unknown. Larger animals suffocate at or below 10 percent oxygen. Yet for millions of years Gaia's energetic disequilibrium and seeming "watchfulness" has preserved a safe and biologically critical level of oxygen.

As anomalous as the high oxygen concentration in Earth's atmosphere is the phenomenally low level of carbon dioxide, CO_2—roughly 0.03 percent. Contrast this with the approximately 98 percent concentration of CO_2 in the atmosphere of Venus and 96 percent for Mars. Just the right amount of carbon dioxide has given Earth its Goldilocks quality of being "neither too hot nor too cold, but just right." Since its thermonuclear fusion fires ignited, the Sun has grown progressively brighter, perhaps by as much as 30 percent during the last four billion years, yet since life began more than 3.5 billion years ago, geologic evidence and the survival of life suggest that the average global temperature has always been in the range 5° to 25° C. In ways that are still not fully understood, life has undoubtedly helped to draw down the level of CO_2, thus maintaining clement temperatures. Microscopic life in the oceans called coccolithophores are collectively the ultimate sink for a massive quantity of CO_2, which transfers to them as water washes into the ocean underground bicarbonate decay products of vegetation. The calcium carbonate exoskeletons of the coccolithophores—the solid embodiment of CO_2—fall as sediment in the ocean. Thus arose the white cliffs of Dover in England. Later, volcanoes at the edge of the Earth's moving tectonic plates on which the continents ride recycle this form of carbon back to the atmosphere as CO_2.

Gaia's network of biospheric control mechanisms seem to blossom whenever a new path presents itself. The atmosphere's storehouse of stable nitrogen gas that mercifully dilutes oxygen and prevents global conflagration is derived from so-called denitrifying bacteria in the soil. Lightning bolts and nitrogen *fixing* bacteria return nitrogen in oxidized form to the land where it can be incorporated in life. At least half of the mass of all terrestrial life lives in the sea where a safe concentration of salt is provided by complex Gaian regulation. The elements iodine and sulfur, which are essential to life, are wafted from the sea to the land in the form of methyl iodide and dimethyl sulfide. Sea life is essential to land life and vice versa—a balance of precisely controlled tension between the two realms.

Large land animals, including ourselves until we acquired significant technology, may be the least important factors in the Gaian scheme. The bacterial world with its endless inventive genetic sharing and recombination is the sturdier force, ready at a moment's notice to redesign itself to meet a new ecological crisis. But in a very real sense multicellular life is still part of that bacterial world because prokaryotes first coalesced to form eukaryotic cells and then those evolved to the complex cellular communities of large-bodied creatures, both plants and animals. But a new life force has come into the world since evolution produced intelligent hominids. Gaia now harbors technology-wielding life forms with the capacity to alter the biosphere for better or worse.

On the negative side, technological civilization appears on the verge of radically altering global climate as it burns fossil fuels and wantonly lofts billions of tons of CO_2 into the atmosphere each year, adding insulation to nature's greenhouse. Scientists expect the CO_2 concentration to double within the next 100 years, and even if it doesn't, significant temperature increases may be on the way. CO_2 is already 20 percent above preindustrial levels and rising rapidly. Mathematical climate models indicate that rising CO_2 and the other greenhouse gases will cause the Earth's average temperature to increase 1.5 to 4 degrees Celsius within the next fifty years, resulting in a global climate significantly warmer than it has been in the past 100,000 years. Unchecked, this will cause major changes in precipitation and resulting agricultural disruption. Even without the melting of land-lying Antarctic and Greenland ice, the warming of the oceans will cause them to expand and their levels to rise. Coastal regions around the world will be inundated, and civilization will have only its own lack of cohesion and foresight to blame. Climate change, though far from a catastrophic extinction of all life, certainly portends a dramatic change in human life-styles.

One possible salutary effect of global climate change may have been overlooked. Observing that human activities had brought about environmental changes on a global scale, now embattled civilizations might be lured from conflict to cooperation essential to avoid more severe climate modification. We have evolved enough intelligence to recognize and anticipate global problems such as the greenhouse threat and have a chance to add intelligent control to the Gaian mechanisms already in place. If we fail to avoid undesirable turns of events in global ecology, Gaia's marching troops—the bacteria—will perhaps come to our rescue with yet another invention. More audacious and unlikely as it now seems, with new techniques of genetic engineering we might design our own "regulator" bacteria to infect the planet and set things right.

If human beings had never evolved and begun the present industrial modification of the atmosphere, large plants and tiny microorganisms would have been undisturbed for millions of years in their active role of reducing greenhouse gases like CO_2. But the levels of greenhouse gases are already percentage-wise so low that as the Sun gradually becomes brighter over the next tens of millions of years, Gaia's natural mechanisms may be hard-pressed to keep the planet cool. Fortunately, human beings increasingly become the mind of Gaia and have the power to resume control. James Lovelock speculates that the highly evolved human instinct for beauty may be a subtle force for planetary survival. He writes, "If the product of all this cooperative effort, a human being, seems beautiful when correctly and expertly assembled, is it too much to suggest that we may recognize by the same instinct the beauty and fittingness of an environment created by an assembly of creatures, including man, and by other forms of life? Where every aspect pleases, and man, accepting his role as a partner in Gaia, need not be vile."

We live in a germinating age of population instability, the threat of nuclear warfare, the possibility of nuclear winter, and environmental pollution crises on a global scale. Any one of these problems could ultimately be lethal to many forms of life, but very unlikely to *all* life. Gaia is sturdier than that. She has endured many previous crises and will probably survive many more. Lynn Margulis says, "I think that Gaia will persist for a long time. But what may look like a minor perturbation from a distance will be exactly the minor perturbation that leads to the loss of what we consider most dear to us, namely, an environment suitable for human habitation."

The idea of a superorganism like Gaia may seem paradoxical to biologists who obsessively insist that an exact analogue of Darwinian evolution must characterize an evolving process. Since Gaia the superbeing has no visible "competitors" in the game of life, how has she managed to evolve? Who are Gaia's predators, and does she have sexual partners? The answer is, of course, none in either department. Gaia embodies a different mode of change, what we might call an emergent evolution that may be the inevitable result of natural selection among myriad life-forms within a biosphere. Though her name was born of mythology, Gaia is no myth. She lives and may soon give birth to a daughter on a nearby world, Mars.

8
Life Beneath Fear and Terror

The metamorphosis of old Mars into a living, terraformed planet would be more than a metaphor of the ultimate conquest of the god Mars' influence in all terrestrial civilizations, or of the victory of life over death which has been the spark behind so much human aspiration. The spread of Earth-born life beyond the world of its biological origin would be an event of galactic significance, both for what would still lie ahead of a newborn multiplanetary human civilization, and for what would be left behind.

—James Oberg, *Mission to Mars,* 1982

Life is everywhere: slithering with the snake through nodding reeds, threading the parched desert with the kangaroo rat, swimming with the ameba in a drop of rain. Even if we project our musings beyond the world, life quickens the planets, binding them without a rope to moons, to suns, to the Pleiades . . .

—Guy Murchie, *The Seven Mysteries of Life,* 1978

The dreams of centuries will soon be fulfilled when intelligent beings walk the deserts of the fourth planet from the Sun. They may be the first Martians, and—in a most extraordinary reversal of classic science fiction—they will be us. Exploration of Mars by people, long the fondest hope of space visionaries and calculating engineers, will probably be attempted within the next twenty-five years by Americans, Russians, groups of other space-faring nations, or all together. This will be the first step in the inevitable colonization of the most clement world in the solar system for us apart from Earth. Will the colonists forever be content to inhabit sterile cocoons as they venture across the rugged surface, or will the yearning to remake Mars in Earth's image become irresistible? Will people someday bask lightly clothed on the beaches of a new Earth—a radically transformed Mars? Whatever the distant future may hold, exploring Mars will mark the beginning—or perhaps the rebirth—of a new planetary superorganism, the first child of Gaia. Perhaps we shall come to know her as "Marsia." (Mars is also known by its Greek name, Ares. Were we to call the new Mars "Area," geometric confusion would surely reign.)

"Terraforming" planets to be habitable by Earthlings has long been a staple of science fiction stories. The word was first used by Jack Williamson in two science fiction novels in the 1940s, but the concept is as old as Utopia, which is, in fact, the name of the Martian desert plain where one Viking spacecraft landed. Yet within the last two decades the idea has acquired a new respectability as planetary scientists, biologists, and engineers have begun to study terraforming or "planetary engineering" in detail. While some of these studies may have been treated at first as mere thought experiments to exercise scientific cunning, their conclusions increasingly suggest the real prospect of altering the most opportune world, Mars, starting sometime in the next two centuries. There are many uncertainties, not the least of which are ethical and social questions.

An irate reader may be forgiven the reaction: "Don't we have enough problems with Earth's own environment without charging off to inflict our ignorance on other worlds?" This view fails to note that by understanding the dynamics of other ecospheres we can hope to fathom the workings of our own Gaian biosphere. In imagining the process by which we could transform a planetary atmosphere to the needs of terrestrial life, we come to grips with natural and cultural forces that may alter our own world for better or for worse. Human civilization may decide never to remake another planet, but it will have learned much about the intricate network of cycles that have sustained life on Earth for billions of years in the face of innumerable catastrophes.

If we do begin to remake worlds, why should the first one be Mars? Astronomers for a long time thought that cloud-shrouded Venus might be a habitable twin of Earth, blessed perhaps with tropical rain forests and graced with life-sustaining oceans. In the early 1960s the U.S. Mariner-2 spacecraft that darted past Venus pulled the veil off the second planet from the Sun. It revealed that erstwhile Aphrodite was, in fact, Dante's Inferno. Temperatures on the Venus surface are over 460° C, more than enough to melt lead, even though Venus is roughly only 30 percent closer to the Sun and is about the same size as Earth. Moreover, any living intruder unwise enough to descend to the surface would be crushed by an atmosphere, predominantly carbon dioxide, with about 100 times the sea level pressure on Earth.

Inklings of Venus's hellish nature came even before this first successful space mission to a planet. In a 1961 article in *Science*, Carl Sagan had outlined the pre-Mariner conception of Venus and suggested hypothetical steps for Venus terraforming. Venus's upper atmosphere might be seeded, he wrote, with a particular strain of microscopic algae that would transform carbon dioxide to oxygen and residues of carbon that would fall to the surface. Rains would come, the surface would cool, and over hundreds of years the hot greenhouse of an atmosphere would moderate in temperature and pressure, paving the way for life on the surface.

That suggestion may have misjudged the mind-boggling difficulty of transforming Venus. Not only were the pressures and temperatures later found to be greater than anticipated, but astronomers discovered that Venus has a day equivalent to

243 Earth days. For a planet with a year of 225 Earth days, this meant that one face was oriented toward the Sun for a substantial period. In this extreme condition, atmospheric circulation could be nothing like that of Earth. Airless Mercury revolves in its orbit closer to the Sun than Venus and is an even less likely candidate for terraforming by foreseeable technology. It too has a day that is a substantial fraction of its year, 59 and 88 Earth days respectively.

In the distant future Earth's Moon may glisten with oceans and cloud cover, but these will have been imported from other bodies in the solar system, perhaps comets or asteroids. The Moon, with a surface gravity only one-sixth that of Earth, will forever be a weakling in holding an atmosphere—speeding molecules would rather leap into space than linger over Selene.

The so-called gas giant planets of the outer solar system, Jupiter, Saturn, Uranus, and Neptune, are nothing like the terrestrial dust motes nearer the Sun. The giants have surfaces that would resemble nothing familiar; slushy swamps or oceans of liquefied gases would best describe them. Even with familiar landlike surfaces, the high atmospheric pressure would crush humans. The warming light of the Sun grows ever dimmer as we journey outward. For each factor of two increase in distance there is a fourfold reduction in solar radiation intensity. Distant and tiny Pluto may be the gateway to the stars, but it is a dark and inhospitable outpost, not a likely world on which to plant an oasis.

There are many substantial rocky moons surrounding the gas giant planets, and some of these might be turned into new Earths. But far more appealing for the near term is the only reasonably warm world that is eminently accessible with current space technology—Mars. Throughout history Mars has been a magnet for the imagination of scientists, dreamers, and poets. Superficially, it is a dry planet with a thin atmosphere and is only half the diameter of Earth. The surface of Mars prophetically equals the combined area of the terrestrial continents. Two irregular potato-shaped moons, Phobos and Deimos (meaning "Fear" and "Terror"), that are conceivably captured wayward asteroids, circle Mars in orbits of 7.7 and 30.3 hours respectively. These telescopically nearly invisible moonlets, 22 and 12 kilometers long, were unknown until American astronomer Asaph Hall observed them in 1877 and named them after Mars's mythical attendants, Fear and Terror. In his fictional *Voyage to Laputa*, Jonathan Swift 157 years earlier had presciently written about two moons of Mars with orbital periods of 10 and 21.5 hours!

It was also in 1877 that Italian astronomer Giovanni Schiaparelli described apparently linear features or "grooves" on Mars that he called *canali*. This led Bostonian astronomer Percival Lowell to posit the now famous irrigation canals originating at the Martian polar caps and crisscrossing Mars—the supposed planetary engineering works of desperate Martians bent on bringing water to arid regions. Belief in intelligent Martians and their canals died slowly, and it took imagery sent back by American spacecraft beginning in 1965 to prove that, in fact, the telescopic observations were the result of wishful thinking that combined faint images of unconnected surface features. But the canals of the mind curiously anticipated possible efforts to change Mars on a planetary scale.

Mars is Earth-like in a number of ways already. The Martian day is only a fraction of an hour longer than Earth's and the present tilt of its rotation axis to the planet's orbit plane, 25 degrees, approximates Earth's season-producing tilt of 23.5 degrees. Mars orbits the Sun in 687 Earth days but on an elliptical path more eccentric than Earth's. While Earth's distance to the Sun varies only between 147 and 153 million kilometers over its year, Mars swings from 206 to 250 million kilometers away from the Sun, producing greater seasonal temperature variation than on Earth. Both planets have northern and southern polar ice caps, though in the case of Mars their extent varies widely over the year. On Mars, the southern cap appears to be solid carbon dioxide or "dry ice," while the northern polar cap is mostly water ice.

The surface gravity on Mars is a comfortable one-third Earth gravity, enough to hold a substantial atmosphere over geologic time. Unfortunately, the surface atmospheric pressure on Mars is a tiny 0.6 percent of Earth surface pressure and is 96 percent carbon dioxide, 2.6 percent nitrogen, less than 2 percent argon, a minuscule 0.03 percent water vapor, and almost no oxygen. Earth's atmosphere is almost 80 percent nitrogen with the balance as oxygen and only minor amounts of other gases. Life on Earth is protected from killing solar ultraviolet radiation by a high-altitude layer of ozone gas (molecules with three atoms of oxygen), while Mars lacks this shield. Martian surface temperatures range from −143° C to occasional balmy mid-latitude highs of 27° C, with an average temperature 60° C lower than Earth—a very cold world, to be sure. In all probability liquid water cannot exist anywhere on the planet.

The greatest difference between Mars and Earth is the lack of or extreme sparsity of life on Mars. The two Viking spacecraft that tested Martian soil for life in 1976, produced inconclusive results at best, although a majority of scientists agree that Viking showed Mars to be, at least currently, a fairly "dead" world. Yet some have interpreted the results positively and suggest that the experiments confirm that Mars has microbial life. Others contend that the experiments, ambiguous or not, could never be considered conclusive, since only two spots were tested. Investigators selected the landing points for flight safety, so they were not the most likely sites for life.

Some have suggested that the periphery of the melting polar caps would be more suitable locations to hunt. Life may hide under some yet unturned stone or even *in* stones as a sort of lethargic "cryptolife," much as microbes that have been found living near the surface of Antarctic rocks, energized by the faint energy of sunlight filtering through the translucent minerals. Or perhaps life arose on Mars, evolved to an unknown pinnacle, and then perished. One of the strongest reasons for detailed exploration of Mars is to search for the fossil evidence of such extinct life. In the absence of definite proof of life, science must ask why Earth and Mars seem to differ in this fundamental respect. At one time Mars clearly had abundant water flowing on its surface and must have been more like Earth. Gullies and erosional features seen all over the planet by orbiting spacecraft are compelling evidence for ancient waters on Mars.

While Venus is trapped in an infernal state, and Mars appears locked in a glacial mold, Earth enjoys an apparently precarious existence, literally between these worlds. As we learn more about Gaian biospheric dynamics, it becomes more apparent that life confers long-time stability to planetary atmospheres. Gaian feedback control mechanisms have kept the atmosphere, land, and oceans habitable even in the face of major external hazards like changes in solar brightness or the atmospheric clouding caused by large impacting asteroids.

Perhaps any serious attempt to terraform Mars would have to employ life-forms on a planet-wide scale, first to generate an oxygen atmosphere and then to insure its stability. This was the conclusion of the first scientific study of Martian terraforming conducted in 1975, before the Viking missions. A group of biologists and atmospheric scientists led by Robert MacElroy and Melvin Averner with the support of NASA and Stanford University, published its conclusions in a landmark NASA report entitled "On the Habitability of Mars: An Approach to Planetary Ecosynthesis." While admitting that there were areas of uncertainty, the study concluded that, "No fundamental, insuperable limitation to the ability of Mars to support terrestrial life has been unequivocally identified."

According to most proposals, terraforming Mars would begin with efforts to increase atmospheric pressure and temperature. The polar caps would play key roles, since if they could be melted, the carbon dioxide and water vapor released would initiate a self-generating greenhouse heating. Some proposals suggest darkening the caps with finely dispersed Martian soil to increase solar energy absorption. Others involve gigantic gossamer-thin orbital mirrors to focus sunlight on the ice. Calculations show that enough gas could arise from polar melting to create in only 100 years a carbon dioxide pressure equivalent to that of Earth at sea level. Depending on how much water could be obtained from the poles—a gap in current knowledge—temperatures might also rise substantially, because water vapor is a much more efficient greenhouse gas than carbon dioxide. Increasing atmospheric pressure on Mars would permit heat to flow by *advection* from lower latitudes to polar regions, and cause further ice melting.

The great hope for Mars is the permafrost which many scientists assume exists below the surface over much of the planet. Vast reserves of frozen water are almost certainly present, given the great geological evidence on the surface. Hopefully, there might also be biologically critical reserves of frozen nitrogen. The warming trend set in motion by polar ice melting could help to free some of these trapped volatiles, leading in many thousand years to large basins of free water. Moderate temperatures and tolerable pressures could allow people to move about the surface, protectively clothed against intense ultraviolet radiation and wearing oxygen masks, but otherwise free to roam the land.

But to get oxygen into the Martian atmosphere will require the services of untold quadrillions of beings. Much as Earth was transformed starting two billion years ago by oxygen-producing bacteria, so the Martian carbon dioxide might be broken down in the presence of water to oxygen and carbohydrates. Terraforming studies point toward the use of specially suited algae or lichens to complete the Martian transformation.

Existing bacteria and lichens might actually be able to live on Mars even now, yet the efficiency of transforming carbon dioxide would not be great. (Biochemist and Viking experimenter Gilbert Levin believes that patches of what he interpreted as greenish color on Martian rocks viewed by a lander are in fact life-forms akin to lichens! But his colleagues disagree.) The 1975 NASA study concluded, "Mechanisms of genetic engineering currently available or under development could be used to construct organisms far better adapted to grow on Mars than any present terrestrial organism." A process that might ordinarily take tens of thousands of years could be substantially reduced by genetically redesigned life. Even very efficient terraforming on Mars may require millennia, a span clearly incompatible with current myopic political time frames. But the first stage—elevating the pressure and initiating the warming trend—could be amazingly rapid.

Other would-be Mars terraformers are not content with the slow path to remake Mars. James Oberg, author of *New Earths*, suggests that diverted asteroids could be used to impact the Martian surface and create depressions 8 to 13 kilometers deep, which would immediately acquire Earth-like atmospheric pressures—a kind of localized terraforming. Oberg's "air holes" are matched in their daring by his "iceteroids," kidnapped fragments of Saturnian moons rich with frozen volatiles sent onto collision courses with Mars to enrich its air. This might be a partial re-enactment of one theory of how volatiles first arrived on the terrestrial planets: the collision of icy comets, rather than the more conventional scenario of gases and liquids "outgassing" from planet interiors. Others have even suggested radical techniques, such as reactivating Martian volcanoes using thermonuclear explosives, expecting them to spew beneficial gases over the planet. The costs of these planetary engineering proposals are, of course, enormous, but who can say what humanity would not do to gain another world?

Accepted geophysical theory correlates Earth's ice ages with the gentle gravitational influences of other planets and the Sun on the equatorially bulging world. So-called Milankovich cycles in climate stem from Earth's orbital eccentricity changing slightly with a period of about 90,000 years, and the polar axis bobbing from a 22- to 25-degree tilt over a 42,000 year period. Mars is thought to have such cycles, though in much exaggerated form. Its polar axis tilts dramatically from 15 to 35 degrees to its orbit over 25,000 years, causing periodically much greater polar warming. The Martian orbit varies 10 to 15 percent in eccentricity over very long periods. How lucky Earth is to have a large Moon, which through its own counterbalancing gravitational torque, damps these complex climate-changing motions. In 1973, scientists Joseph Burns and Martin Harwit suggested that an asteroid suitably placed into orbit around Mars could stabilize polar wobbling and preserve the hard-won gains of terraforming.

If even microscopic life-forms are found on Mars, many scientists would consider upsetting the Martian ecology to be unethical. The genocide of indigenous Martian life to suit our own purposes would certainly be scientifically unconscionable and morally repugnant. But what if transforming Mars were to enhance such life? How long could we be content with the role of zookeepers for Martian life?

Other scientists who are strong advocates of space colonization scoff at terraforming, favoring instead colonies consisting of large rotating cylinders or spheres freely floating in the depths of space. Physicist Gerard O'Neill envisions ecologies clinging to the inner walls of these island earths, pressed against the walls by the artificial gravity of rotation—an environment precisely tailored to human needs, a culture built entirely from accessible asteroid or lunar resources and solar energy. The inner surface areas of these space colonies may ultimately be many times the area of Mars, and coming and going to them, spacecraft would not have to fight the gravity of a large planet.

No doubt, terraforming and free-floating colonies will have their place in the expansion of life into the cosmos. Russian space visionary Konstantin Tsiolkovksy predicted 100 years ago that people would not live in the "cradle of Earth" forever. As significant an evolutionary leap as life's origin may be the taming of other worlds—manifest destiny extended to the heavens. Meanwhile back on Earth, the "Mars Underground" community of scientists and engineers struggles to make the first step to Mars a reality. Wernher von Braun in 1953 drew up the most detailed plans for a massive manned expedition to Mars in *The Mars Project*. Many other Mars mission engineering studies were done in the 1960s. The 1981 "Case for Mars" report describes the results of a conference held in Boulder, Colorado, to plan for an early Mars base with perhaps a dozen explorers. Since in situ production of food, biologically essential gases, and rocket fuels is possible on Mars, the favored philosophy is to establish a permanent colony at the outset to which crews would rotate on shifts lasting from two to five years.

Many analysts agree that initial Mars missions could be conducted for one-third to two-thirds the cost (in fixed dollars) of the Apollo Moon project. Rapidly advancing space technology eliminates the need to reinvent many wheels.

The earliest human expeditions to Mars may in fact be to its outer moon, Deimos, an excellent vantage point from which to observe the planet. Explorers could repetitively send dozens of unmanned rovers down to the surface to hunt for life and to return geologic samples that would be analyzed in a well-equipped Deimos laboratory. An explorer on Deimos with the aid of television cameras could "drive" a rover on Mars as though he were in it, unhampered by the long signal delay if he controlled it from distant Earth. In their own right,

Deimos and Phobos may even be ideal worldlets to colonize permanently.

It may not be for exclusively pragmatic reasons that we go to Mars and change it forever, bringing with us interplanetary spores, the genetic heritage of Gaia. We may go for aesthetic or psychological purposes through the energies of dreamers and misfits—a course not unknown in history. An outpost of terrestrial life on Mars would be the beginning of an insurance policy for the continuity of civilization and the delicate web of Gaian inventions. Planetary disasters—impacting asteroids, nuclear winters, and environmental troubles unforeseen—would then be far less likely to silence the last human voice.

9
THE BIOTROPIC UNIVERSE

If we are examples of anything in the cosmos, it is probably of magnificent mediocrity.

—ERIC CHAISSON, *Cosmic Dawn*, 1981

To a worm in horseradish, the whole world is horseradish.

—Old Yiddish saying, quoted by RABBI KUSHNER,
When Everything You've Ever Wanted Isn't Enough, 1986

Were we gifted with the vision of the whole Universe of life, we would not see it as a desert sparsely populated with identical plants which can survive only in rare specialized niches. Instead, we would envision something closer to a botanical garden, with countless species, each thriving in its own setting.

—GERALD FEINBERG AND ROBERT SHAPIRO,
Life Beyond Earth, 1980

The Earth spawned life and a Gaian aura that later blossomed into conscious awareness. We are life and are surrounded by life, yet when challenged to abstract the most universal qualities of what it means "to live," the wisest seers fail to agree. If we hold up a mirror and say, "This life," can we be sure that other life, looking into a similar glass while struggling to understand itself, would include us in *its* universal definition? Without a satisfactory definition of life, how can we think about testing the waters of the cosmic ocean in search of life "as we don't know it"? We have already sampled the Moon and a bit of Mars, but before going much farther, we

Adapted from a 1902 sketch by G. Melies titled *Trip to the Moon*.

ought to pause and re-examine our prejudices about what we will admit into the "house of life."

Life Unbounded

We could adopt the parochial view that life must incorporate elaborate fantasies of carbon atoms—organic molecules—and must always appear in the form of cells or cell colonies like our metazoan selves. Shrinking our imaginations ever further, we could insist that everything in the universe entitled to the name "life" must bear the indelible mark of the nucleic acids, DNA or RNA. Yet as chauvinistic as we might be about carbon atoms and the virtues of the genetic common carriers of terrestrial life, we instinctively know that these characters could be far from universal attributes of things that we might quickly recognize as being alive. It may be that anything that would ever tempt us to call life is made of carbon atoms, but we can't shake the gnawing suspicion that this is a provincial delusion.

Perhaps exotic chemistries we can barely imagine are the essence of the living kingdom on other planets. Maybe the universe teems with alien chemical life, but the carbonaceous variety is freakish. Or does extraterrestrial life sometimes partake of physical phenomena and exist in environments that seem impossibly antithetical to life—the throbbing surfaces of stars, for example, or the nearly empty reaches of interstellar space? Do we perhaps inhabit, for want of a better term, a *biotropic* universe—a cosmos that delights in casting up startling varieties of organization that are indisputably "lifelike" but that manifest themselves in realms far beyond the experience of any one breed of generic "life"? If we were to discover that the universe teems with chemical life, carbon-based or not, it would joyously confirm the vision of a quickening universe. But, oh, how much more awesome, to be immersed in a sea of biotropic organization—in the hearts of stars, deep inside planets, with forms almost as incomprehensible to ourselves as we to ants?

There is a quandary about what are *necessary* and what are *sufficient* conditions in a definition of life. The distinction is important, because if a phenomenon possesses only attributes necessary to be called life, it still might not be life. An example of a necessary condition is that life certainly must be "complex" in some ill-defined sense. No one would suggest that a volume of pure water was alive in anything but a questionable metaphysical sense; it is just too "simple" an entity, an endless repetition of the same basic molecular subunit. (Yet if we were ignorant of atomic theory and chemistry, we might believe that the sloshing vibrancy of jiggling water made it seem alive!) We would have more trouble with a magnificent enclave of crystals. A tour through a natural history museum's beautiful display of crystalline minerals, bulging with variegated colors, shapes, and sizes, never fails to impress with its uncanny resemblance to a botanical display. And if we had watched those crystals grow by means of time-lapse photography, their vitality would have astonished us. But crystals as *we* know them, though they may be complex and even be (as we discovered in Chapter 6) the possible precursors of life, aren't in our lexicon of the living. But on another world, crystals may have acquired additional properties through evolution that may have given them lifelike qualities.

Complexity may be in the eye of the beholder, and we must admit that to a unified planetary scale intelligence evolved to Gaian dimensions, we mortals may seem to be uninteresting, simple cogs. For us, an operational definition of complexity, without waxing mathematical, may be the quality of having a significant *functional* intricacy on distance scales very small compared to overall size. In other words, life must have hierarchical structure of subfunctions. From this perspective, a cell would have a degree of complexity necessary for life, as would we intricate beings who are composed of from 10^{15} to 10^{16} exquisitely linked cells. And a Gaian superorganism would be a life-form without question. A conventional crystal, with myriad intricate atomic inclusions and fault networks, would be perhaps adequately complex to be called life, yet its subunits lack significant hierarchies with adequate functional intricacy.

Another necessary quality for life is that it must have evolved. Life without evolution—not necessarily Darwinian evolution—is patently unimaginable. Of course from the wide perspective of *cosmic* evolution, everything, even atoms themselves or rocks, may be said to have evolved. But the kind of evolution that must be inseparable from what we choose to call life is a process that can in theory be traced back in time to ever more fundamental levels of simplicity. In that sense, evolution and complexity are virtually identical *necessary* qualities of life, one being dynamic and the other static. Atoms and rocks have the necessary property of having evolved, but the stages through which they emerged seem so conceptually simple that we can't consider them alive.

There are many attributes of life that have ocasionally been considered necessary but upon more reflection turn out not to be. One is the ability to self-reproduce sexually or asexually. A mule, the infertile child of fecund parents, is stubbornly alive, yet it cannot reproduce. This is a peculiar counterexample, but nonetheless adequate to put one on guard about elevating mere reproduction to an essential property of all life in the cosmos. In more general terms, we might imagine a single organism evolving to gargantuan size by attaching ever more intricate subunits to itself in response to environmental change. Evolution for such a unitary being would clearly not be Darwinian, although the entity's subunits might evolve by a kind of natural selection. Gaia resembles a creature some of whose subunits (classical life forms) evolve in the Darwinian fashion and other components by physical self-regulation or "cybernetically," akin to thermostatic control. But there may exist other classes of unitary beings that don't need reproducing components.

Though reproduction may not be necessary for life "as we don't know it," self-repair is. Life must have mechanisms for internal reconstitution after the inevitable degradation that comes from any environment's multiple insults. Even in a very benign setting there will be unavoidable disruptions to any form of life as elementary particles and radiations course through the universe and penetrate the most sheltered Shangri-La. When terrestrial life forms lose their self-correcting ability, death is the result. A perfect mechanism for self-repair means immortality for life so blessed. Immortal life-forms may even be more common in the universe than imperfectly self-repairing mortal beings. Life doesn't need perfect self-repair, but without some reconstituting ability it may never leave the womb of jostling matter and energy.

Nor is mobility a necessary feature of life. Most familiar life-forms do move, whether it is the imperceptible climb of blades of grass toward the Sun, the wiggling tail of a sperm cell seeking an egg, or the charge of a cheetah after its prey. Yet what do we make of the wind-borne spore, quaking not at all within its solid sheath? It is life, as is the virus, immobile itself but pregnant with potential for an encounter with a chemical factory—a cell, whose machinery it commandeers for its reproductive benefit. The remarkable tardigrade, a microscopic bug, can remain dehydrated and immobile for years, stirred into activity only when doused with water.

If complexity, evolution, and self-repair are necessary for life, and reproduction and motion are not, what then is *sufficient* to define life? Now we are on much shakier ground. After all, computers and any indubitably complex artifacts made by people certainly sit at the top of an incredible evolutionary pyramid that includes all of terrestrial life. And on the scale of the tiny crystalline pathways in a micro chip, computers are functionally complex, but are they alive? We are justified in answering no, even as we contemplate the day when computers will have exceeded the powers of human intelligence over a broad range of abilities, much as some computers already have done over a narrower range. Computers are evolved and complex, but they aren't alive. Not yet, though someday they may have earned the right to the distinction.

The more we let our imaginations roam, the more we realize that there are probably *no sufficient conditions* adequate to define life, qualities which if possessed would automatically confer the title. There are only the *necessary* conditions of complexity, evolution, and self-repair. Science fiction and daydreaming about alien beings may be as appropriate a guide to possible life as rigorous definition by illusory sufficient conditions. Phenomena, qualified by necessary conditions, we should decide to call life with that intuitive recognition of the living that only life itself is heir to.

Strange Life

Can we be certain that "life" cannot exist in radically different environments? One hint that it can is that on Earth carbon-based life already has adapted to seemingly lethal conditions. Microbes live happily in intensely salty brine pools, highly acid or alkaline solutions, boiling hot springs, bubbling sulfurous mud pots, and pitch-black thermal vents on the ocean floor. Only a small leap of the imagination brings us to life based on radically different chemistry that accommodates to even more peculiar conditions. Life that doesn't use water as a solvent is one example. The most often cited nonaqueous liquid medium for alien biochemistry is ammonia (a four-atom molecule with one nitrogen and three hydrogens), gaseous at "room temperature" but liquid at scores of degrees Celsius below zero. Chemists have speculated that on the surface of a very cold planet, swimming in a sea of ammonia, there could exist complex molecules incorporating linked nitrogen atoms rather than carbons. Life with an affinity for ammonia instead of water might be the common biotropic heritage of deadly cold moons and planets far from their parent star. Does nitrogenous life frolic on fantastic ammonia seas within our own solar system? Voyages to the moons of the outer solar system may someday tell.

Lower temperatures generally make chemical reactions proceed much more sluggishly, leading doubters to believe that "cryonic" or low-temperature life, dissolved in ammonia or another liquid, would not have had time to evolve, although, given much more opportunity than the present age of the universe, it could. But our grasp of chemistry at low temperatures is so limited, experiments being much more difficult at temperatures where gases liquefy and even freeze, that one may doubt but not entirely dismiss the possibility of cryo-life.

Perhaps, as Gerald Feinberg and Robert Shapiro suggest in *Life Beyond Earth*, organized complexity in deep cold could appear in physical rather than chemical form. On a planet orbiting a dim white dwarf star or far from a red giant, an ocean of liquid hydrogen could be home to strange amorphous entities of solid hydrogen, buoyed by entrapped bubbles of helium. The tiny beings, no larger than small protein molecules, would draw energetic sustenance from weak stellar radiation that would create changing physical states in different zones within them. Constellations of icy hydrogen "cells" might evolve into larger creatures. If beings of the cold abyss grew to perceive the wider cosmos, they might have less belief and comprehension of the rare element carbon-life in warmer oases than we have of them in their frigid torpor.

What about life on planets with searing hot surface temperatures, or inside a world like our own planet, which has a core temperature exceeding 6,000° C—hotter than the surface of the sun? Some scientists have proposed that the element silicon, from the same chemical family as the ambidextrous carbon atom, could be the basis of what we might call "magmatic life." If there were a star with the brightness of the Sun and a planet, much closer to this star than Mercury is to the Sun, its surface might bubble with lava that never solidifies. In past centuries some astronomers reported seeing such a world orbiting the Sun, which they called Vulcan, whose existence has never been confirmed and is unlikely. Within its molten "oceans" a silaceous form of organization might arise. Silicates, like the mineral quartz, are crystal networks of silicon atoms each bound to four oxygens with other "contaminant" atoms substituting now and then for a silicon. At temperatures over 1,000° C, the rigid crystal pattern dissolves to a fluid broth of silicate chains. Magmatic creatures might bask on a starlit planetary surface, or swim in the radioactive depths of a molten planetary interior, occasionally ignobly coughed up by a volcano and suffering a death of terminal solidity. Their skeletal outlines might be fossilized in rocks picked up by unsuspecting carbon-structured surface explorers.

Nor should we imagine that the surfaces and interiors of planets and satellites are the only habitats for life. The stars themselves could harbor organization, not fragile chemical life, of course, in the hot, gaseous interiors of stars like the Sun. In typical hot stellar innards atoms lose their grip on their outermost electrons, the resulting broth of partially clad nuclei and electrons being called a plasma. The ionized gas is energized by nuclear fusion reactions occurring deep within the star. Intense magnetic fields twist and warp filaments of plasma, occasionally squirting it out into space in huge arcing prominences. It would seem that no structures could survive the apparent plasma chaos. But a few scientists have speculated otherwise, and have suggested that stellar plasma could evolve living domains of persisting magnetic organization—almost literally sunspots but on a much smaller scale than those huge enduring magnetic "storms" visible on the Sun's surface. If glowing stars truly can be abodes of life, perhaps even intelligent life, there would be new meaning to the ancients' deification of the Sun and other bright stars.

Not all stars are globes of plasma. There are relatively cool brown dwarf stars with solid surfaces, more akin to bloated planets with many times the mass of giant Jupiter, equivalent to 318 Earths. It is possible that thin films of chemical organization may hug some of these little dark worlds, undetected sites for the evolution of life that may even be more numerous than their brightly shining sisters.

Astronomer Frank Drake suggested that life processes could be based on reacting nuclear particles—protons and neutrons—and inhabit the surface of superdense stars called neutron stars, the cinders of titanic supernovae explosions. On the surface of such stars that often rotate hundreds of times per second, the density may reach ten or more tons per cubic centimeter. Instead of chemical reactions, neutron star beings would embody interactions of nuclear particles which are millions of times faster. Evolution of neutron star life would occur with blinding speed. The microscopic creatures would be unlike any with which we are familiar, yet physicist Robert L. Forward has chronicled the evolution on an entire civilization of neutron star beings, the Cheela, in his imaginative science fiction novels, *Dragon's Egg* and *Starquake*.

In the vast reaches of interstellar space there might be entities composed of extremely tenuous matter—interstellar dust grains, molecules, and atoms organized with electromagnetic fields. Astronomer Sir Fred Hoyle first popularized the idea of living interstellar nebulae in his science fiction novel *The Black Cloud*. These electromagnetic entities would be millions of kilometers across. To recharge their unimaginable metabolic cycles they might covet occasional encounters with drifting stars. If nebulous interstellar monsters possessed intelligence, attempting to communicate with them, possibly via radio signals, would be literally like talking into space.

Intelligent nebulosities don't exhaust the imaginative possibilities for life and complex organization in the cosmos. Consider the intricate spiral structure of a wheeling galaxy. Could its hundreds of billions of stars be mere cells or atoms within the body of a "living" island universe, a kind of "gravitational life" on a stupendous scale? Stars within a spiral galaxy take tens or hundreds of millions of years to orbit the galactic center, which is often home to a massive black hole that gobbles up stars by the millions. If some galaxies "live," the time scale of their metabolism or "thoughts" would certainly be reckoned in billions of years.

Consider the invisible, so-called dark or strange matter that measurements show must permeate a spinning galaxy to make its glowing stars behave as they do. Cosmologists have yet to reveal its subnuclear identity. Could this strange matter organize itself into living structures interpenetrating ordinary matter wherever it has congealed into galaxies, stars, and planets? Strange matter would seem to be the strangest life of all in the bestiary of potential cosmic organization. But the list of possibilities for life is endless in the universe's garden of cosmic delights, and we, like proverbial ants, may be oblivious to a world of organization above our noses.

Returning from an untethered flight of imagination, let us not forget more mundane niches for life in the outer solar system that beckons modern-day Magellans. The peculiar chemical coloration of the atmospheres of the giant planets with warmer temperatures deep within them may be crying out for us to explore and discover strange new life-forms flying or floating in their swirling clouds. Saturn's moon, Titan, the only satellite in the solar system with a substantial atmosphere, probably has an extensive ocean of liquid hydrocarbon compounds. Jupiter's moon, Europa, is an ice-sheathed world that is likely to have a hidden ocean of water a few hundred meters below its surface of cracked ice. What life may lurk in these deeps?

Genes, Memes, and Technology

The biotropic universe may be wishful, imaginative fiction or the actual plan of a cosmos that is rife with complex organization. Whether or not life can manifest itself in more than one basic way, it may ordinarily start as an experiment in organic chemistry, evolve intelligence, and then shrug off its mortal coil of carbon imprisonment. As a "hobby"—a mere evolutionary experiment—life may then invent another form of chemical or physical organization to replace its outmoded design. Glimmerings of that prospect are occurring on Earth at this very moment with the advent of human-designed electronic computers. It seems likely that "silicon intelligence" will remain safely under our watchful eyes for the foreseeable future, but perhaps someday people will give up organic infirmity and pass their heritage of learning and passion down to stronger, cooler intellects.

It is difficult enough trying to fathom the biological evolutionary process that led to us, without attempting to imagine how alien life would evolve, what inheritance code it would incorporate, and what means of selection would rule. But we may guess that alien life, whatever its biotropic form, would pass through at least three major stages in evolution. In the first or genetic stage there is chemical or physical evolution—selection of basic morphology, and behavior. This is the stage of terrestrial evolution in which "selfish" genes (so characterized by biologist Richard Dawkins) regulate the form and behavior of "gene survival machines," i.e., biological bodies. Dawkins' view is that genes, not individuals, act as though they were the fundamental units for whose "benefit" Darwinian natural selection is occurring. The idea is entirely reasonable if the metaphor of "selfishness" is not taken too literally. Darwinian evolution operates statistically through the success or failure of individuals, not with literal conscious desires on the part of DNA molecules. Even though Lamarckian evolution, transformation by characteristics *acquired* during an individual life, appears to play little if any role here on Earth in the first stage of evolution (the symbiosis of primitive cells may be an exception), it is conceivable that alien life would evolve by a different selection process, such as a Lamarckian one.

Human beings are now in the second stage of evolution—the cultural or *memetic* phase. Richard Dawkins coined the term *meme* to refer to cultural replicators that are "capable of being transmitted from one brain to another." Memes are ideas, fashions, art, etc.—entities that evolve extragenetically and extrasomatically. Memes presuppose brains or analogous organs to manipulate them in the intricate process of memetic evolution. In a sense, memes are the prisoners of genes, for we cannot memetically generate that for which our brains have no capacity. For example, the average person couldn't compose a symphony as appealing to fellow human beings as Beethoven's 9th if he or she struggled a lifetime, much less attempt the wondrous feat in fifteen minutes. But cultural evolution is fantastically more rapid and creative than biological evolution. Genetic engineering and biotechnology, themselves memes, are now even making possible the alteration and potential major redirection of genetic evolution. Thus, we arrive at the third stage of evolution: technology.

Technology is essentially an extension of memetic evolution, but has the potential to become an evolving entity itself, as biologist Lynn Margulis and her son Dorion Sagan suggest in *Microcosmos*. Primate technology, from printing presses to space shuttles and nuclear weapons, is still under human control—relatively speaking. Human brains are now required to coax it along and govern its evolutionary path. But technology is now palpably accelerating and directing itself, quickening in the dynamic sense now, but potentially quickening in the sense of coming to life. Computers now design parts of computers; machines are beginning to design and fabricate other machines.

In the 1950s mathematician John von Neumann proved that machines could be built to construct replicas of themselves. In recent years there have been vigorous discussions of using such replicating robots in space exploration. The latter day von Neumann machines would seek out and mine the resources of the Moon and asteroids for human consumption. But before long, intelligent machines might beget mechanisms that would be increasingly more talented than the originals. The new children of humanity, imbued with the passions of love and learning, might become our symbiotic counterparts, or our lineal descendants, if we decided to hand down the stewardship of terrestrial life. If naked primitives can already comprehend the grand drama of tripartite evolution happening here on Earth—genes, memes, and technology—perhaps we should throw caution to the wind in imagining the biotropic wonders of deep space and still deeper time.

10

WHISPERS IN THE ZOO

We cannot be kept in a zoo forever without becoming at least dimly aware of our predicament. . . .

The idea that we shall be welcomed as new members into the galactic community is as unlikely as the idea that oysters will be welcomed as new members into the human community. We're probably not even edible.

—JOHN BALL, 1980

Despite the utter mediocrity of our position in space and time, it is occasionally asserted with no sense of irony, that our intelligence and technology are unparalleled in the history of the cosmos. . . .

We think it possible that the Milky Way Galaxy is teeming with civilizations as far beyond our level of advance as we are beyond the ants, and paying us about as much attention as we pay to the ants.

—CARL SAGAN AND WILLIAM I. NEWMAN, 1983

Like a swarm of fireflies blinking on and off on a warm summer night, galactic civilizations may dawn and then disappear. One may flash brilliantly but not persist long enough for any other civilization to recognize its existence. If the brief light of a single firefly represents the duration of one community in the galaxy, it is mere chance that others will be lit simultaneously. But during its lifetime the galaxy of 400 billion stars may give birth to millions of civilizations, and so increase the chances that many will overlap in time. The more frequently civilizations arise, the shorter the average physical distance between them.

The analogy of fireflies isn't entirely accurate, however. A firefly can flash many times in one place, but a civilization may have only one opportunity. If a culture destroys itself or is snuffed out by natural causes, it is gone forever unless organisms surviving on the world evolve and replace it. A firefly becomes aware of a flashing neighbor virtually instantaneously, because the speed of light is so great and the range to the nearest neighbor so short. Galactic civilizations may be widely dispersed and have such a fleeting existence that any external evidence of a culture's presence, whether or not intended, dissipates before the community plays out its short role. Fireflies also have a mutually understood signaling system that was developed by evolution. The channels through which radically different galactic civilizations may have contact are almost certainly not so well established.

Though the analogue of the firefly flash—interstellar communication with radio waves or other radiation—may be a practical and inexpensive way to bridge the interstellar gulf, direct contact via space flight will have a special lure for the *first* civilizations to emerge in a galaxy. The early risers will have few or no partners with whom to engage in remote conversations. Challenged by the silence of space, the intrepid early explorers may roam their galaxy seeking mute, primitive life-forms. And like fireflies in a wooded habitat, civilizations that have mastered interstellar travel may by colonization spread offspring throughout the starry environment. Moreover, since our Milky Way galaxy is about 10 billion years old and the solar system less than half that, it is possible other civilizations began a program of interstellar migration long ago. Why, then, do we not see evidence of alien visitation all around us, since interstellar travel should have been available to extraterrestrials for millions if not billions of years?

This apparent quandary has been called, apocryphally, the Fermi Paradox. About 1950 at Los Alamos, after an informal discussion about travel between the stars, it is said that the great nuclear physicist Enrico Fermi posed the impertinent question, "Where is everybody?" Since then, dozens of essays that address this alleged mystery have been written. But is Fermi's question really paradoxical? Astronomer William I. Newman believes to the contrary—if the question is cast in a more complete form, namely:

> *If* extraterrestrial intelligent life is abundant and *if* space travel is relatively easy and *if* advanced civilizations will feel "compelled" to explore the galaxy and can do so successfully, and *if* they have had enough time to do so, and *if* we have tried hard enough to find them, then shouldn't we see evidence of extraterrestrial life?

Whether evidence of an extraterrestrial presence in the solar system passes beneath our noses unrecognized is highly debatable. But if the evidence exists, it certainly isn't speaking loudly or coherently. On the other hand, some serious investigators of the UFO phenomenon say that the evidence of extraterrestrial visitation *is* persuasive. Equally serious UFO debunkers believe otherwise. True, there is a residue of some highly unusual, unexplained UFO cases that rises above the murk of blatant hoaxes and misperceived planets and aircraft. But useful as such research may be, it has not yet presented the necessary *incontrovertible* evidence that unexplained UFO reports indicate a bizarre extraterrestrial intervention. Extreme skepticism is essential in dealing with this slippery subject, but we should be open-minded and not dismiss all claims that UFOs may be the answer to Fermi's question.

As for the abundance of galactic civilizations, we are either alone in the universe or we are not. If a barren universe—a condition of mind-boggling cosmic infertility—seems intellectually and emotionally impossible to accept, how prevalent do we imagine life and intelligence to be? This question becomes an elaborate scientific parlor game. We cleverly craft "estimates" based on educated guesses of the frequency of suitable planets (or other imagined sites), the probability of life arising at each viable location, the probability that intelligence will emerge in that life, etc. Then, to estimate the number of currently existing civilizations, the combined probability factors for sites, life, and intelligence must be multiplied by an estimate of the average longevity of extraterrestrial societies. Applying all our knowledge of cosmic evolution, we find probably not *billions* of independently evolved civilizations in the Milky Way, but perhaps *millions*. Of course there may be far fewer or we may be alone, but if millions of independently evolved civilizations coexist, then a typical society may well have been around for millions of years.

When contemplating the typical lifespan of alien cultures at a particular level of advancement, we are really speculating about their biological and cultural development. And this will remain speculation unless we can tap into a galactic exobiological survey conducted by another civilization—the mythical *Encylopaedia Galactica* envisioned by Carl Sagan in *Cosmos*. A *Catch 22* situation: we won't have a firm idea of the prevalence of extraterrestrials until we discover them—or communicate with others who have. We can't reliably estimate how hard they are to detect until we detect at least one extraterrestrial civilization!

Starflight

Their motivation may be questioned, but the *ability* of advanced extraterrestrials to travel between the stars can't be doubted, because we young upstarts have already discovered, in theory, many general approaches to interstellar flight. The next horizon has always challenged human beings, and the interstellar gulf is no exception. Would-be galactic sailors—physicists and engineers—have identified a number of possible propulsion technologies and projected optimal trajectories to various stars.

The central difficulty, of course, is the vastness of the distances to be traversed. The nearest destination for us, Proxima Centauri, is more than 250,000 times farther from the Earth than the Sun—a journey of 4.3 years as light flies. Today's chemically propelled spacecraft, or even nuclear rockets that are already possible given the motivation to build them, would take tens of thousands of years to get there. Thus, robot probes will probably be our first interstellar emissaries. In fact, several probes are already under way—the U.S. Pioneer-10 spacecraft crossed the outer bounds of the solar system in July 1983 while still sending back information. Voyager I after its swing past Saturn in 1981 also technically became starbound, and Voyager II will do the same after its close encounter with Neptune in 1989.

If we are wise and ambitious enough, more capable missions may depart around the year 2000, sending back data for up to 50 years and from 1,000 times farther out from the Sun than Earth. They would probably use refined versions of propulsion systems now being developed, such as the "ion drive" engine—electric propulsion. This would use electricity generated by a nuclear reactor to transform propellant into charged particles—ions—which would then be accelerated by an electric field and expelled at high velocity to provide efficient thrust.

Even more ambitious probes driven by advanced propulsion systems might depart by the middle of the next century. For example, in the mid-1970s the British Interplanetary Society conducted perhaps the most detailed feasibility study of an interstellar voyage. In the society's scenario, a starship, named Daedalus after the legendary inventor in Greek mythology, would make a flyby mission to Barnard's star located 5.9 light years away. Daedalus would be propelled by thermonuclear "microexplosions" ignited by laser or electron beams within millimeter-size fuel pellets 250 times a second. The ship would accelerate to about 12 percent the speed of light—some 36,000 kilometers per second, or more than 1,000 times faster than today's rockets. The one-way trip time: only 50 years.

"Light sails" are another possibility for starships of the twenty-first century. Manufactured in space, the sails will be made of thin metallic film and will be hundreds of kilometers across. They will tow payloads tethered to the sail by thin, ultrastrong cables. Some of the sails will be propelled by radiation from the Sun—or someday by radiation from other stars. Powerful lasers stationed in orbit around the Sun may push other sails out of the solar system.

Planetary and environmental scientist Gregory Matloff and I have determined that a sail can be blown out of the solar system by sunlight at more than 1 percent of the speed of light after passing relatively close to the Sun. Thus, small robot probes towed by reflective sails might reach Proxima Centauri in only 350 years. But gaining true freedom in the cosmos may require leaps of science and technology at the frontiers of our imagination. Glimmerings may already be seen. For example, there is the "interstellar fusion ramjet" proposed by physicist Robert Bussard in the 1960s. A starship a few kilometers long generates a huge magnetic field that collects interstellar hydrogen to burn in its fusion engines. The craft would collect fuel faster as it speeds up, which in turn would make the ramjet work even better.

Physicist Robert L. Forward of Hughes Research Laboratories and others have investigated the production, storage, and use of antimatter for rocket propulsion. In such an engine matter and antimatter would come together and instantly annihilate each other, creating a flash of energetic gamma rays. The most promising antimatter rocket would use protons and antiprotons, because along with gamma rays their reaction produces charged particles called *pions* that can be funneled through a magnetic nozzle to create thrust. The antimatter rockets common in science fiction may be distant, but their lower-performance cousins might well be used in the next century for exploratory flight into interstellar space.

The question of human travel to the stars reflects something of what we believe extraterrestrials might do. For a trip to Epsilon Eridani, a nearby Sun-like star 10.8 light years away, Robert Forward proposes a 3,000-ton human habitat towed by a light sail 1,000 kilometers in diameter. An array of solar-powered lasers orbiting the planet Mercury would send a beam of light to a thin plastic focusing lens, also 1,000 kilometers in diameter, stationed between the orbits of Saturn and Uranus. The powerful beam would accelerate the sail and payload to one-half the speed of light in 1.6 years. The probe would then coast until the destination neared. Forward suggests an ingenious plan to decelerate the probe. It requires the sail's 320-kilometer diameter mid-section, coupled to the payload, to detach near the end of the mission and act as a retro-reflector of light from the remaining outer ring that accelerates ahead. The one-way journey to Epsilon Eridani would take a mere 20 years, but the power consumption during acceleration and deceleration, though obtained from "free" sunlight, would be a fantastic 43,000 times current world power usage.

The possibility of humans traveling even faster at high "relativistic velocities" is intriguing because of the practical benefits of relativity's "time dilation." This makes flight of almost any distance possible within a human lifetime, if—a mighty big if!—speeds of more than, say, 99 percent of light can be reached. As a starship approaches light speed, its mass increases according to relativity, and the energy used to achieve even small increases in velocity becomes enormous, wastefully inefficient. For this reason relativistic manned spaceflight must be viewed as a very distant prospect. But advanced extraterrestrials may have far more energy at their command for such extravagant uses. And a human lifespan, so short compared to the time required for "slow" starflight, might not loom as a major deterrent to beings who live much longer.

If humans can't travel fast enough to take advantage of time dilation and can't find some way to extend lifespan artificially or stop life temporarily (perhaps suspended animation by controlled freezing), they might instead travel in multigenerational "interstellar arks." As early as 1929, British crystallographer John Desmond Bernal envisioned space colonies journeying through the galaxy in starships, or even in hollowed-out asteroids. Today designs for orbiting space colonies promoted by physicist Gerard O'Neill and others make interstellar arks seem more feasible. Nay-sayers scoff at the notion of colonists setting out on journeys that only their descendants will complete. But the idea of traveling for millennia may be less intimidating to people who have lived their lives in a rotating cylinder the size of a small city that orbits the Sun.

If we who are only a few bare centuries into the era of technology can propose feasible though costly plans for interstellar travel, what should we expect of cultures incomparably more advanced and wealthy? Perhaps the best answer is Arthur C. Clarke's wise "Third Law": "Any sufficiently advanced technology is indistinguishable from magic." After all, human civilization has developed its own magic. Television would surely have baffled Isaac Newton, and a man-sized cylinder that can destroy an entire city would have astonished anyone prior to 1945—and most of us since. It is indeed reasonable to assume that the forbidding interstellar distances are not impossible barriers to highly motivated explorers among the stars.

The Greening of the Galaxy

It has been said with much truth that if the dinosaurs had been smart enough and had wanted to colonize the Milky Way, by now they could have done so. Alas, endowed with but meager intelligence, they perished. How, then, can we reconcile the apparent absence of extraterrestrials near Earth with a quickening universe of abundant civilizations? A number of valiant efforts have addressed this question mathematically by modeling how fast a civilization would spread through the galaxy if it decided to undertake interstellar migration. Some of the earliest estimates in the 1970s concluded that colonization of the entire galaxy by a single civilization could be extremely rapid—from a few million to 10 million years. The process is analogous to molecules of colored dye diffusing outward from an ink-drop into a pool of water. An even more exact analogy is the way species of animals, including humans, have radiated and migrated across the globe. Unless some strong universal constraint prevented *every* one of millions of civilizations from pushing galactic migration to the limit, there should be an extraterrestrial presence here, now, and everywhere.

Taking the argument to its illogical conclusion, a few physicists have cited the apparent local absence of extraterrestrial technology as evidence that humanity is alone in the cosmos. Physicist Frank Tipler, notably, has imagined self-replicating "von Neumann machine" probes cast into space by aliens—robots with advanced artificial intelligence that use extraterrestrial materials to replicate like bacteria. Their mission: to explore ever-increasing territory in the galaxy. If we see no evidence of this "intelligently" launched technological cancer of machines that "eat" planets, certainly the original extraterrestrials who dispatched the probes never existed or else died out early in their development. Such a plague certainly could devour the galaxy in a relatively short time, but it is hard to believe that a single civilization even out of millions would attempt such a potentially suicidal act. Even if it were tried, other societies more rationally disposed would undoubtedly oppose the impolite cosmic madness.

More recent and refined calculations by Carl Sagan and

William I. Newman have shown that if a single extraterrestrial civilization sets out to colonize the galaxy, its progress may be much slower than earlier estimates—requiring hundreds of millions or even a billion years. The exact times depend on the details of migration, but are always much more significant fractions of the lifetime of the galaxy than a few million years. Thus, extraterrestrials from a single originating society would not necessarily be here. On the other hand, there may be many extraterrestrial civilizations. What powerful inhibition prevents *all* of them from spreading throughout the galaxy? Here intuition fails us and we can only guess. The psychology of a very advanced civilization may simply be incomprehensible.

There may be many reasons why the behavior of advanced civilizations would violate the ordinary mathematical rules of ecological expansion common in animal and human migrations. Perhaps colonization has been tried often, but deliberately and invariably remains highly restrained. There may be an inevitable turning inward among advanced cultures. With age and intellectual and emotional contentment, they may no longer seek new physical frontiers. Perhaps civilizations of "philosopher kings" have tasted enough of the real universe after a few cautious journeys. They may then stay close to home and play hedonistically with "computer" simulations of the evolution of alien worlds and beings—like our movies, only with much higher fidelity. With uncharacteristically anthropocentric reasoning, Carl Sagan and William I. Newman suggest that there may be an "intrinsic instability of societies devoted to aggressive galactic imperialism." Loosely translated, "In the long run bad guys can't win because they destroy themselves."

Perhaps the simulations of galactic colonization err by incorrectly assuming that populations of *individual* beings multiply and diffuse, leaping off one world in search of another as soon as the local ecosystem has been saturated. Unitary beings, akin to integrated planetary organisms like Gaia or

Marsia, may be more common biological units in the galaxy. Smaller intelligent beings living within them may be merely the "genes" of Gaian entities. It may be more appropriate to develop and apply rules of Gaian diffusion—if such could be conceived— than conventional equations of biological migration. The expansion of unitary beings would perhaps be more robust and enduring, but slower than the spread of weak "bacteria" like ourselves.

Harvard University radio astronomer John Ball believes we may live in a cosmic "zoo" and that this may explain the absence of aliens. Advanced cultures do not violate our territory, either because they fear rude and dangerous adolescent behavior or they simply want to observe pristine biocultural evolution without disturbing it. In another variant of the "Zoo Hypothesis," Frank Drake, a pioneer in the radio search for extraterrestrial civilizations, suggests that aliens may frequently acquire biological immortality. This may make them extremely safety conscious and afraid of contacting "hazardous" young races on distant planets. The immortals may prefer to hide among the stars, occasionally dispatching robot probes to interfere with and redirect dangerous evolutionary developments on other worlds.

The "Zoo Hypothesis" has a strangely alluring logic. Ironically, because who wants to be on display in a zoo? Yet for whatever reasons extraterrestrials may have imposed their quarantine, it explains the multidimensional "silence" that we increasingly recognize. Other intelligences may be profoundly aware that the universe is everywhere quickening, and may act in their own interests with great circumspection. Astrophysicist Glen David Brin writes, "It might turn out that the Great Silence is like that of a child's nursery, wherein adults speak softly, lest they disturb the infant's extravagant and colourful time of dreaming."

Perhaps our zoo or nursery is deliberately leaky. Now and then extraterrestrials offer tidbits of information borne by electromagnetic radiation or messenger probes. We can hope that advanced civilizations have not grown thoroughly introverted and irrationally fearful of the animals in their transparent cages. Thus, a search may be in order for whispers of gentle encouragement to our struggling infant culture.

We stare at the fertile stars and their silence haunts us. What cool intelligences lurk behind those fires we can only dream. What secrets do they whisper and to whom? We are a weak race with lofty ambitions, pretenders to an invisible galactic empire that may be only an ephemeral vision. Our mighty starships that set sail for worlds beyond the Sun are, for now, only spaceships of the mind. Perhaps the time has come to cease our foolish counting of civilizations on pinheads, like medieval philosophers divining angels. As our childhood comes to an end, we must listen intently for whispers in the zoo.

Adapted from a photograph of the Project Ozma antenna at the National Radio Astronomy Observatory, Greenbank, West Virginia.

11
FROM OZMA TO META

With our scurrying minds
and our lidless will
and our lank, floppy bodies
and our galloping yens
and our deep, cosmic loneliness
and our starboard hearts
where love careens,
we are listening,
the small bipeds
with the giant dreams.

—DIANE ACKERMAN, "We Are Listening," 1986

This may be the least important story of the day . . . or it may be the most important story of our time.

—CHARLES OSGOOD, *CBS Evening News, March 7, 1983*
Comments at the dedication of Project Sentinel

The priceless benefits of knowledge and experience that will come from interstellar contact should not come too easily. To appreciate them, we should be required to devote a substantial portion of our resources, our assets, our intellectual vigor, and our patience.

—FRANK DRAKE, "On Hands and Knees in Search of Elysium," 1976

In 1900 the French Academy of Sciences offered 100,000 francs—the Pierre Guzman Prize—to the first person establishing communication with a world other than Mars. So certain was the presence of Martians at the time that communication with them seemed too easy to deserve a prize.

The scientific view of the prospects for contacting extraterrestrials has matured enormously since that quaint era of perceived Martian canals and presumed Martians. We are now nearly certain that there is no other technologically advanced civilization in our solar system, although Boston University astronomer Michael Papagiannis and a few others believe that aliens may now secretly inhabit bodies in the asteroid belt between Mars and Jupiter. Experiments by the Viking landers in 1976 at best hint at the possibility of microbial life on Mars—and those elaborately conceived tests may have more prosaic explanations, such as unusual inorganic chemical reactions that mimic metabolic processes. The irrefutable discovery of even microbial Martian life, evolved independently from life on Earth, would point to the likely plenitude of organisms in the universe at large. But while arguments may continue about whether microbes exist on Mars, humankind must look to the 100 billion or more stars in our own Milky Way galaxy or to the billions of galaxies beyond, if we are truly to contact our biological or biotropic kin.

Since the Milky Way is at least 10 billion years old and our solar system less than five billion years, civilizations elsewhere could be millions of years older than our own. Would the wisdom of age cause them literally to shine their beneficence upon us via radiations that we could someday detect? "It may be that if we just point our antenna at the right piece of sky, and tune to the right channel, we will discover signals giving us knowledge that other beings have been storing for aeons," says astrophysicist Thomas McDonough of the Planetary Society. "Even if we could not decipher them, just knowing of one other civilization advanced enough to broadcast signals could be vital." This key point is often absent in popular conceptions of interstellar communication. Receipt of one such signal would show that an extraterrestrial culture could overcome the problems of its youth and survive to old age.

Many optimists in the search for extraterrestrial intelligence (SETI) feel that a renaissance is at hand. In their view, a persistent search with radio telescopes combined with sophisticated yet economical microcomputer-controlled signal processing might well lead to detection of ETI signals within the next fifty years. New listening efforts are under way, attempts which go technologically far beyond earlier tries. An international SETI petition in 1982 signed by 73 scientific notables has urged "the organization of a coordinated, worldwide, and systematic search for extraterrestrial intelligence."

Astronomer Frank Drake, a SETI pioneer in the 1960s, says, "The greatest obstacle to the success of radio searches has not been the lack of sensitivity to weak signals or too narrow coverage of the sky, as most imagine. Rather, it has been our inability to study, at any given instant, a large sampling of frequency channels of the most promising portion of the radio spectrum." However, the new computerized signal processors have allowed scientists to search first on tens of thousands and now on millions of channels simultaneously. This is a vast increase in ability to search large portions of the frequency spectrum.

The current renaissance in SETI has been brought about by the parallel development of inexpensive microprocessor electronics and very low noise-level receivers of great frequency bandwidth and tunability. The improved computer technology permits almost instantaneous recognition of signals as noise or message, so vast quantities of data need not be stored for later automated or manual processing. Current SETI programs project a 10-millionfold or greater increase over past searches in the frequencies and directions to be searched in the next decade.

Ozma and Beyond

In the first realistic attempt to detect extraterrestrial signals, called Project Ozma, in 1960, Frank Drake used an 85-foot diameter dish at the National Radio Astronomy Observatory in Green Bank, West Virginia, to monitor two nearby sunlike stars, Tau Ceti and Epsilon Eridani. Drake named the project after the princess in the land of Oz, "a place very far away, difficult to reach, and populated by exotic beings." Scientists in several nations have since made more than three dozen other searches. For example, Robert Dixon of Ohio State University has for more than 10 years scanned the sky using a radio telescope the size of a football field. (Rescued by outraged supporters from conversion to part of a golf course, "Big Ear" searches on.) Researchers at the Jet Propulsion Laboratory and the University of California at Berkeley have "sneaked" observing time, using antennas usually employed to track interplanetary spacecraft. The piggyback nature of the project earned it the name *Serendip*, also an acronym for Search for Extraterrestrial Radio Emissions from Nearby Developed Intelligent Populations. And in the Soviet Union V. S. Troitsky and colleagues at the Gorky Radiophysical Institute have scanned the skies since 1970 for sporadic pulses of possibly intelligent origin. The vitality of the once essentially nonexistent SETI field might be judged from a 1980 bibliography by myself, Robert L. Forward, and others that listed almost 3,000 technical and popular references.

Although none of these projects has yielded unambiguous evidence of alien transmission, there have been some exciting moments along the way, such as signals later found to be terrestrial interference, and a curious few that were fleeting and did not persist long enough to be identified. Perhaps the most famous false alarm came in 1968 in England, at a radio telescope not even dedicated to SETI, when Jocelyn Bell, a Cambridge University graduate student, and Anthony Hewish, her professor, detected regularly spaced pulses arriving every one and one-third seconds. It turned out that they had detected pulsars, the neutron stars that had been predicted by physicists. But at least for a while Bell and Hewish referred to their signal as "LGM-1," for Little Green Men.

Contrary to popular belief, all previous searches have not been extensive enough to have had a significant probability of success in a sparsely populated galaxy. "The searches to date have been like trying to find a needle in a haystack by walking past the haystack every now and then," says Frank Drake. Indeed, only a small fraction of the "cosmic haystack" has been surveyed, and there is so much territory that previous part-time and even full-time efforts would be unlikely to have borne fruit. Some of the searches targeted hundreds of specific stars in the solar neighborhood that were deemed likely to have life-bearing planets. These stars were examined extensively at many frequencies. Other searches have been "whole sky," sweeping large areas of the heavens without focusing on specific nearby targets but monitoring a smaller part of the radio spectrum. The number of stars searched so far, says Cornell astronomer Carl Sagan, constitute "only a millionth of a percent of the stars in the galaxy."

Today's radio telescopes could detect beams directed at Earth from identically sized and sufficiently powered antennas located on the other side of the galaxy. Of course, if we were to try to detect "leakage" signals from advanced civilizations that were not deliberately beamed at Earth, similar to our radio, radar, and television signals now leaking into space, antennas would need to be much larger than any now available, and costs would rise dramatically. You need big ears to detect a faint noise. And much as terrestrial civilization increasingly puts its communications in shielded pathways (cables, optical fibers, etc.) and thereby reduces the inventory of leakage signals, extraterrestrials might similarly do so early in their communications development. Nothing is lost for the moment, however, in neglecting to search for leakage signals that may no longer be prevalent. We have to start somewhere.

The idea of monitoring radio waves rather than other regions of the electromagnetic spectrum, such as visible light, infrared, or X-rays, traces back to 1959. Philip Morrison and

Giuseppe Cocconi, physicists then at Cornell, suggested in *Nature* magazine the advantages of searching radio frequencies. Specifically, they proposed that the microwave region was best suited for SETI—the frequency range between about 1 billion and 10 billion hertz, or 1 to 10 gigahertz. (A hertz is one cycle per second.)

Why microwaves? This region is a relatively quiet "window" for listening. The galaxy is a noisy place. Cosmic static includes noise generated by fast-moving charged particles in interstellar space, as well as remnant echos of the birth of the universe—the Big Bang. Radio telescopes themselves have so-called internal quantum noise, and molecules in the Earth's atmosphere absorb some incoming radiation. In all, the microwave region is where the noise is least and reception is best. Thus, an extraterrestrial transmitter beaming microwaves wouldn't have to "shout" so loud to be heard—a fact that radio astronomers everywhere in the galaxy would know.

But that is still an enormous range of frequencies. What is needed is some universal frequency marker that would be recognized by civilizations that had not previously communicated. Morrison and Cocconi solved this problem as well, suggesting that one particular frequency within the microwave window might be a "magic frequency." The idea is that this "magic frequency" would be a signpost or probable meeting place of galactic cultures in the electromagnetic spectrum. As an earthly example, if you are trying to meet a friend in New York City but have no idea of the meeting site, you might reasonably hover around major points of interest such as the Empire State Building or Times Square, hoping your friend will also have that idea.

Morrison and Cocconi suggested a magic frequency of 1.420 gigahertz, or 1,420 megahertz, as it's more commonly called. Radio waves of this frequency with a corresponding wavelength of 21 centimeters are the ubiquitous natural emissions of neutral hydrogen atoms, the most abundant element in the universe. The emission is caused by the electron of the hydrogen atom spontaneously changing its direction of spin. Fortuitously, this is also in a part of the spectrum that is particularly low in interference from other natural sky radiations.

The Morrison–Cocconi proposal has withstood the test of time: radio astronomers still consider 1,420 megahertz fertile ground for SETI. And researchers have also proposed other potentially interesting frequencies in the microwave region. For example, hydroxyl (OH), a molecule composed of one atom of hydrogen and one of oxygen, emits a number of distinct frequencies near 1,666 megahertz, just up the radio spectrum from neutral hydrogen. Since hydrogen and hydroxyl are the products of the decomposition of water, the frequencies near those they emit have been romantically dubbed the "water hole." SETI researchers suggest that galactic species might mingle in this region via radio waves, just as animals congregate at water holes on Earth.

If one examines all potential radiative channels of interstellar communication, the logic of using electromagnetic waves in the microwave region is even more compelling. If beams of charged particles, such as electrons, protons, or ions, are considered, the dispersion and curving of such particles by galactic magnetic fields becomes an immediate deterrent. The beaming of massive neutral (uncharged) particles such as atoms would be ruled out by the extreme expense of accelerating them to near light speed, the speed at which all electromagnetic radiation travels. As for neutrinos, speculative faster-than-light tachyons, or radiations unknown, they would be difficult to modulate and receive (at least by foreseeable technology). At present we are powerless to do anything with these beams of science fiction.

With electromagnetic waves, radiation that is not significantly altered by the interstellar medium over long distances, we find a curious and fortunate minimum in the natural background radiation. In most regions of the spectrum, interference makes it more difficult to recognize artificial signals that may be generated by extraterrestrials. Even though an advanced civilization might command vast energy resources, it still might be guided by a "principle of economy" in attempting to attract the attention of less capable societies. The portion of the spectrum that maximizes the probability of their signal's detection at the lowest energy expenditure to them appears to be the "water hole" and adjacent frequencies. The aliens simply have to do less "talking above the noise." So the "water hole" is also a depression or "hole" in the electromagnetic listening spectrum.

There is still the difficult problem of finding the "needle" signal in the "cosmic haystack." This can be conceived as a "volume" with frequency as one dimension, antenna pointing direction as the second dimension, and receiver range from the alien transmitter as the third dimension.* Only a small fraction of the cosmic haystack has been searched in prior listening efforts, so it is not surprising that no confirmed alien signals have been detected. There is so much ground to be surveyed that past part-time and even full-time efforts would be unlikely to have met success. The problem is not so much one of sensitivity in receivers, because receivers are already sensitive enough (except for "leakage" searches), but of persistence in exploring the nooks and crannies of the cosmic haystack. If the deliberately beamed transmissions are there, we have a good chance of finding them with a dedicated and extensive search using existing radio telescopes.

Sentinel

The new direction in SETI began at the 84-foot-diameter radio telescope in Harvard, Massachusetts, a privately funded

*The third dimension is technically receiver sensitivity, related to detectability of a given power transmitter at various ranges.

effort by the Planetary Society, the Pasadena, California, based international citizens' group in support of space exploration. On March 7, 1983, principal investigator physicist Paul Horowitz of Harvard University, Carl Sagan, Philip Morrison, and other prominent figures in the SETI community ceremoniously launched "Project Sentinel" by anointing the telescope with champagne. The parabolic dish at the Oak Ridge Observatory had been rescued at the eleventh hour from permanent deactivation, like the "Big Ear" rescue, another small victory for the cause of cosmic awareness. When radio telescopes are someday granted the respect they deserve, we'll know that our civilization has outgrown its period of juvenile delinquency.

The giant metallic ear pointed skyward on that gray March day in the idyllic New England town known otherwise for its spectacular apple orchards. Delectable croissant in hand and recalling the cadence of Maurice Sendak's *Where the Wild Things Are*, Carl Sagan exclaimed, "Let the search begin!" as he swung his whole body in the cheer. The champagne bottle showered quintillions of water and alcohol molecules onto the dish as Paul Horowitz's wife, Carol, launched the new journey into the cosmic sea.

The key to the project was an advanced receiving system, a computerized spectral analyzer that handled signal processing and identification. The analyzer scanned more than 131,000 radio channels at once, a tremendous increase over earlier receivers. Horowitz developed the system, dubbed "Suitcase SETI" because of its small size and portability, while on sabbatical at Stanford University in 1981. Colleagues at Stanford and the University of California at Berkeley, computer scientist David Brainard of Harvard, and M.I.T. engineer John Forster also contributed. Horowitz tested Suitcase SETI in May 1982 at a giant, 1,000-foot-diameter radio telescope at Arecibo, Puerto Rico. Looking back, Horowitz says, "That experiment, in which we spent 75 telescope hours looking at 250 nearby stars similar to our sun, constituted the world's most sensitive (though not the most comprehensive) search for extraterrestrial signals thus far."

But Project Sentinel, in round-the-clock operation through mid-1985, was a whole-sky search rather than one focused on specific stars. It thus pointed at millions of stars in the galactic neighborhood. In the first minute of the Sentinel search, the equipment duplicated the equivalent of 1,000 years of cosmic haystack searching by Frank Drake's pioneering Project Ozma equipment that used a parabolic dish with a similar diameter for 150 hours! In its first week Sentinel performed a more extensive search than all previous searches in all nations combined.

Sentinel simultaneously scanned an unprecedented 131,072 microwave frequency channels covering a microwave band only 2,000 hertz wide (in each of two so-called *circular polarizations* of the incoming radiation). Each channel was only about 0.03 hertz in width. The search was the more sensitive in that very narrow frequency channels or "bins" were examined. Calculations show that the distortion of electromagnetic signals during their interstellar journey is only about 0.01 hertz in frequency across a transmission distance of 300 light years, so it is feasible for aliens to send very narrow band signals and expect them to be clear at great distances. The heralding or attention-getting part of an anticipated alien transmission is likely to be very narrow band, both for energy economy and to raise suspicions about the signal's artificial character. There are no known or presently conceivable astrophysical phenomena that would give rise to such narrow band signals as Sentinel was able to detect. The fantastic sensitivity that the 84-foot telescope achieved with these narrow frequency bins allowed it to discriminate a signal of only 0.05 billionth of a watt distributed over half of the Earth—about the energy change of a pinhead falling a distance of two-millionths of a meter onto the earth's surface! (This is equivalent to detecting a twin of the Arecibo radio telescope, 100 light years distant, radiating one megawatt of power.)

Sentinel incorporated a feature that made the search incredibly resistant to false alarms caused by man-made interference, the gremlin and time waster of many past efforts. If suspicious signals are found too frequently, the time taken to re-examine their apparent source to see if they persist can waste valuable search time. Researchers programmed the receiver to adjust continuously for a shift in incoming signal frequency caused by the Earth's rotation and its motion around the Sun—the so-called Doppler shift. The expectation is that if aliens want to beam a message deliberately toward our Sun at a "magic" frequency and have it be the same frequency at Earth, they would adjust for the frequency shift caused by their known relative motion toward the Sun, which they could readily determine by spectral analysis of the Sun's light. The aliens would presume us wise-enough searchers to remove the additional frequency shift due to Earth's motion, which they would be unable to determine. The wonderful byproduct of this strategy is that human-made interference is smeared over a large number of narrow channels as the receiver shifts in frequency or "chirps" (analogous to a bird's chirp) and would not register as a distinct sharp signal.

Each day the telescope beam was pointed at one celestial latitude, and the Earth's rotation caused it to sweep a Moon-sized path completely around the sky. The next day the telescope pointing direction was shifted half a degree in latitude. Thus, it scanned the latitudes from 45 degrees south of the celestial equator to 60 degrees north—about 90 percent of the visible sky—in about 210 days. At that point another "magic" center frequency was chosen and similarly scanned.

Sentinel collected and temporarily stored in computer memory about 30 seconds of incoming signal. It then looked for and recorded the outstanding signals while another 30 seconds of signals were being gathered. Therefore, no observing time was lost. An operator usually scanned the data for any unusual features that the computer selected (those ordinarily sought were signal peaks—the top ten channels from each of the two antenna feeds, one for each so-called circular polarization of the incoming waves). The only thing not fully automated was the daily change in the telescope's celestial latitude setting.

When interesting signals arose, the telescope was repositioned when the target was again visible, and a special tracking program examined the suspected source in detail. A persisting narrowband signal would raise great suspicions, and the world radio astronomical community could then examine its detailed characteristics intensively. For example, a narrow herald signal might "point" to a broader band in the spectrum having more complex form, a signal normally hidden in the background noise.

META

On September 29, 1985, at another emotion-filled dedication at the radio telescope in Harvard, Sentinel grew to 8.4 million channels and became META for Megachannel Extra-Terrestrial Assay. The incredible advances in microcircuit technology since Suitcase SETI was first built in 1981 had brought chip prices down threefold, permitting the almost 100-times expansion of the search for only $100,000. The new bandwidth is 300 kilohertz of high resolution frequency channels. In its first 30 minutes of operation, META searched more of the cosmic haystack than all previous SETI projects combined.

One of META's advantages is that it is able to detect transmissions that are not beamed directly at the Earth, as the present system assumes. In the META scenario aliens can transmit powerful omnidirectional beams without worrying about correcting for frequency shifts caused by relative star motions. The analyzer's expanded bandwidth capacity is large enough to catch a beam even if it has drifted off the original magic frequency.

And there is more—a direct connection between cosmology

and SETI. Philip Morrison of M.I.T., who catalyzed SETI research more than two decades ago, pointed out that META's expanded frequency coverage will let the system take advantage of the so-called cosmic rest frame. The "glow" of microwave radiation left over from the Big Bang of creation is all around us and uniform in every direction. "This is a 'fixed platform' that we cannot ever visit," says Morrison. "But by analyzing that glow, we can—as any astronomer anywhere can—determine our motion relative to that platform."

So what aliens might reasonably do is to transmit many beams at 1,420 megahertz, the magic frequency of hydrogen, but shift each beam they send (depending on its direction) to the frequency it would have if it were in the cosmic rest frame. Thus, each sender and receiver is always able to "correct" for "his" own motion. This, Morrison says, would be "the most magic of magic frequencies," since every advanced civilization would certainly recognize it. The new META analyzer handles a frequency bandwidth large enough to allow for our present error in determining this universal standard of rest. "The scheme promises a meeting place in not too wide a range on the radio dial that can include every would-be communicator within our Local Group of galaxies," says Morrison. "However, galaxies that are really far away still raise problems."

As for government financed programs, the NASA budget has for the past few years included about $1.5 million annually for a projected multi-year SETI program. Always in jeopardy, this funding was acquired after years of neglect due to politically strong skeptics, whose scientific understanding of SETI (or science in general) would have trouble filling a thimble.* The NASA program now under way plans to use a

*Dedicated scientists in touch with the program have formed a nonprofit organization, called the SETI Institute, to furnish labor and other resources to NASA at minimum overhead. Contrast this zeal in husbanding precious resources with much other government contracting, in which inflated overhead expenditures often help to augment profits.

combination of whole-sky surveys and targeted star searches to expand the tiny volume of the cosmic haystack that has been examined.

The key to the NASA effort, as with the META project, is a fast multichannel spectrum analyzer with more than 8 million channels. Work is progressing at NASA Ames Research Center, Stanford University, and the Jet Propulsion Laboratory. The frequency channels will have variable width and will give much broader coverage (1, 32, 74, and 1,024 kilohertz). Using existing radio telescopes and spacecraft tracking antennas, the NASA instrumentation will comb the frequency range of 1.2 to 10 gigahertz as well as select bands between 10 and 25 gigahertz in the whole-sky search. Eight hundred nearby stars (out to 80 light years) and selected distant stellar concentrations, such as star clusters in the Milky Way and distant galaxies, will come under closer scrutiny in the frequency range 1.2 to 3 GHz and selected bands between 3 and 25 GHz. The NASA approach will go significantly beyond META by using elaborate computer programs to hunt for more difficult signals—ones that drift in frequency or are pulsed. The incoming data will exceed a billion bits per second (a bit is a binary digit: a zero or one), and complete analysis of the signal as it comes in will be mandatory. NASA's SETI project in its first hour of operation will exceed all previous searches combined and will ultimately be 10 billion times more extensive than all earlier efforts.

Hope for Large "L"

No one anticipates quick success in finding alien communications because the territory of the search is very large. "You don't do something like this and think you're going to detect a signal next week," says Horowitz. "Our chances of picking up a signal may be small. But if we don't try, they're zero." The greatest uncertainty plaguing SETI theorists lies in two factors that govern estimates of the number of extant communicating civilizations: (1) The fraction of evolved "intelligent" communities that would attempt interstellar communication with less advanced societies; and (2) The longevity of such civilizations in the communicative phase. A rough proportionality exists between the number of extant communicative civilizations in our galaxy, *N*, and their average communicative longevity, *L*. (This was first proposed by Frank Drake in his now-famous "Drake equation," $N = (R_* n_e f_p f_l f_i f_c) \times L$). * Depending on whether one is a SETI optimist or a pessimist, the proportionality is anywhere from an optimistic N equal to L all the way to N being a small fraction of L (with L expressed in years). One of the very great benefits of SETI searches is to gain insight into how large L might be by slowly compiling evidence for an "upper limit" on the prevalence of communicating extraterrestrial communities. The longevity of galactic civilizations is so paramount in SETI that Walter Sullivan in his seminal book *We Are Not Alone* (1964), dedicated it to "those everywhere who seek to make 'L' a large number."

After years of searching, if efforts to listen in the microwave region of the spectrum do not reveal alien transmissions, it will be time to expand the search to other portions of the electromagnetic spectrum. The time will have come to plan larger, more sensitive antennas to listen for leakage transmissions, and perhaps look for faint optical evidence of distant planetary cultures. For example, Princeton University physicist Freeman Dyson has suggested that cultures may have surrounded their parent star with energy-collecting structures that emit

---

*Where R_* = the average number of sun-like stars forming in the galaxy each year
n_e = the average number of clement worlds within a solar system
f_p = the fraction of stars having planets
f_l = the fraction of planets on which life evolves
f_i = the fraction of cases in which life evolves intelligence
f_c = the fraction of intelligent civilizations that attempt interstellar communication.

characteristic infrared radiation. Other scientists have proposed that advanced societies might deliberately seed their star with rare radioactive elements whose spectra would be evidence of technology. The microwave search itself will not have been wasted even if no ETI evidence is found. New tools of radio astronomy will have been developed, perhaps serendipitous astronomical discoveries made, and new information handling technologies created that are certain to find application elsewhere. It is almost a rule in astronomy that whenever a new observational technique is developed, new astronomical discoveries are made.

Of course, we can remain hopeful that extraterrestrials will indeed "phone Earth." The cost of the search is incredibly small compared to its monumental benefits, even without receipt of a bona fide signal. If nothing else, SETI will help to focus humanity's attention on a common purpose—a cause to inspire all but the most wooden psyches. It is noteworthy that the cost of all the SETI programs of all the nations up to now is less than the price of a single military helicopter. A few minutes' worth of the present rate of worldwide military spending would be adequate to support for two decades a major search program with state-of-the art equipment at existing radio telescopes. Or, if even a few minutes of the mounting cost of preparations for global conflict is deemed an indispensable stimulant to political micro-minds, perhaps they could agree to encourage setting aside a few hours' worth of civilization's annual expenditure on cosmetics—thus funding a major SETI program. This seems not an unreasonable payment for experiments that at best might revolutionize human culture, and at least serve as a reminder of our continued quest to know our place in the cosmos. And if a signal *is* found, what music it may be to our ears!

12
SILENT SYMPHONIES

The passion for science and the passion for music are driven by the same desire: to realize beauty in one's vision of the world.

—HEINZ PAGELS, *Perfect Symmetry*, 1985

In the beginning was the absolute rule of the flame: The universe was in limbo. Then after countless eras, the fires slowly abated like the sea at the outgoing tide. Matter awoke and organized itself; the flame gave way to music.

—HUBERT REEVES, *Atoms of Silence*, 1984

Pi is pi, mathematically speaking, whether near the Sun or orbiting the star Tau Ceti. So the argument runs of those who have considered the probable content of radio communications from alien civilizations. Scientists are confident that mathematical concepts could be the basis for a language of interstellar discourse, at least for initial "conversations." Receiving a message indicating, for example, a sequence of indivisible prime numbers (1,2,3,5,7,11,13...) would prove that a signal was from an intelligently directed source. Following that, an expanding array of symbolic mathematical relations and sophisticated concepts would merge easily into language instruction, aided perhaps by pictures and diagrams encoded in the message.

Adapted from a photograph of the Very Large Array, Socorro, New Mexico.

In the nineteenth century, long before radio technology, Carl Friedrich Gauss, who was one of the greatest mathematicians, proposed arousing the attention of Martians by planting huge forests in the shape of three squares touching at their corners—an interplanetary demonstration of the classical Pythagorean theorem. But for more serious interstellar discourse via radio, unless the aliens spoon-feed us with what Philip Morrison calls anticryptological messages (those designed to be decoded by galactic novices), we may find intercepted extraterrestrial broadcasts incomprehensible. Possibly as impenetrable as they might consider our music.

This has not stopped both scientists and theatrical artists from proposing interstellar communication with music as the medium. Recall the melodies employed in Stephen Spielberg's movie *Close Encounters of the Third Kind*. A crescendo of tones accompanied by flashing colored lights establishes communication with a landing party of extraterrestrials. Now traveling to the stars aboard the Voyager I spacecraft (and aboard Voyager II, bound for the stars after an encounter with Neptune in 1989) is a symbolic long-playing copper record that contains 90 minutes of music from many lands and cultures. In the unlikely event that aliens ever come upon this tiny relic, presumably they would possess a terrestrial musicological slice from Bach to obscure tribal songs. But would they appreciate it as "music," or to them would it be discordant noise? Not important, say the scientists and musicians who collaborated in selecting the music. The copper discs were really intended to be "messages to Earth"—a statement about the ideal of harmony among peoples on this now strife-torn planet. Are the foundations of musical appeal to humans, grounded as they appear to be in mathematics, really likely to underlie the psychology of extraterrestrials? This question is clearly in league with, "Is mathematics a universal language?" or "Is consciousness for a dolphin the same as for a human being?"

But perhaps "music *between* the spheres" really *is* part of the usual conversations between members of the galactic club. Could the ultimate evolutionary "purpose" of music be a kind of cosmic social glue connecting distant worlds—much as terrestrial music has the power to transcend national boundaries? And do galactic civilizations at different stages of development appreciate different musical styles, much as our individual musical tastes change throughout life? Maybe whispers of alien music—silent symphonies—are at this moment irradiating the globe, undetected and unrecognized by our finest instruments. Someday we may be smart and patient enough to capture and enjoy celestial melodies if they are there for the hearing. For the moment we are mere Earth-bound babblers inadvertently casting our music onto the waves of the cosmic abyss.

Music looms large as a phenomenon in the evolution of life, not only in the role it might play in interstellar communication. But of course, anything affecting the evolution of life is inseparable from the SETI problem. Music confronts science with a mystery on the grandest scale. On Earth its origins are as obscure as the beginnings of human language to which it may be closely tied, yet it pervades all of nature from birdsong to the cryptic buzz of fruit flies. Ancient Greek philosophers connected the transcendent beauty of musical harmonies with mysterious mathematical ratios of whole numbers that described the most harmonious combinations of vibrating strings. And twentieth-century mathematicians continue to discover new patterns in music's labyrinthian complexity. Above all, music reveals the profound mystery of how we perceive the world and adds to the puzzle of consciousness.

What is music and how did it come to be? Is it native to only one small planet, or do kindred harmonies course through the star lanes of the galaxy? Could we appreciate the music of another world and vice versa? How and why did symphonies of Beethoven and concerti of Bach crawl out of matter at this local plateau of cosmic evolution? Music, like mathematics, seems almost a metaphor for "mystery"—questions about the meaning of music arise faster than they can be answered. There are many more speculative ideas about music than firm conclusions.

Even the most inquisitive might pause to wonder whether answers are even desirable. Douglas Hofstadter, author of *Gödel, Escher, Bach*, wrote in his scintillating book *Metamagical Themas*, "To me, the deepest and most mysterious pattern of all is music, a product of the mind that the mind has not come close to fathoming yet. . . . I don't believe those mysteries will ever be truly cleared up, nor do I wish them to be. . . . I don't wish the fruits of my research to include a mathematical formula for Bach's or Chopin's music. Not that I think it possible. In fact, I think the very idea is nonsense. But even though I find the prospect repugnant, I am greatly attracted by the effort to do as much as possible in that direction."

Music partakes of the same mystery as chemistry and biochemistry—simple atoms, or frequencies for music, combining to create elegant complexity, life, and melody. Music is constituted with sound, but is so much more than a summation of frequencies. Sound is nothing more than waves of compression moving though a medium like air or water. The pressure waves reach human or animal ears, and through exquisitely evolved biological machinery are transduced to electrical impulses that feed into brains. But our emotional response to music goes far beyond a note-by-note recognition of frequency, duration, and amplitude. Indeed, music dissected, heard one note at a time, is obviously no longer music. Yet unlike painting, music can't be appreciated "all at once" the way spatially distributed visual art can. Music is an essentially temporal phenomenon.

The musical experience could be described as a complex reverberation within neuronal networks, highly conditioned by pre-existing activity. Every nuance of one's physical and mental state comes together to transform the temporal acoustic input into higher-level feelings. Physicist David Bohm suggests that music may allow us to experience what he calls an implicate order, a paradigm of the way quantum mechanics makes this an apparently holistic universe—observers are tied inextricably to observed phenomena. In his book *Wholeness and the Implicate Order*, Bohm writes: "At a given moment a certain note is being played but a number of previous notes are 'reverberating' in consciousness. . . . In listening to music, one is directly perceiving an implicate order. Evidently this order is *active* in the sense that it continually flows into emotional, physical, and other responses, that are inseparable from the transformations out of which it is essentially constituted."

Computers have been programmed to compose music by randomizing notes, constraining the tones somewhat with "rules" of harmony and aesthetics. But what music machines currently compose is pallid fare compared to what the human spirit has been able to create. Douglas Hofstadter calls the suggestion that a computerized version of Chopin or Bach may soon be upon us a "grotesque and shameful misestimation of the depth of the human spirit." He says, "A 'program' which could produce music as they did would have to wander around the world on its own, fighting its way through the maze of life and feeling every moment of it. It would have to understand the joy and loneliness of a chilly night wind, the longing for a cherished land, the inaccessibility of a distant town, the heartbreak and regeneration after a human death. . . . Therein and therein only lie the sources of meaning in music." Both Hofstadter and Bohm believe that music involves a transformation of life itself into a temporally experienced pattern. Would the temporal patterns of auditory stimulation that are pleasing to us, derived as they are from a specific cultural milieu, resonate pleasantly in extraterrestrials?

Human musical ability is so wonderful because it seems to

be such an evolutionary "excess." What survival value for our species should we attribute to musical composition in its most sophisticated forms? Of course, many other human abilities are perhaps equally puzzling—did we really need extraordinary mathematical faculties to dominate and surpass now extinct hominid cousins? Human intelligence has been adequate so far to moderate the harshness of nature, but it has elaborated less critical functions seemingly as a byproduct. Still we are drawn to suspect that there was some primitive forerunner of musical ability, most probably human vocal communication.

It is difficult to identify any animal or insect species that doesn't rely in part on acoustic communication. In some of these, the use of sound has risen to a level indisputably musical. Birdsong, for example, is certainly melodic, and we likewise admire the underwater serenades of whales. The clack and babble of all the lives on this planet, if heard all at once, would literally be a deafening cacophony. This terrestrial symphony is driven by mating urges, territoriality, and the need to coordinate group movements. Human beings are removed from much of this biological necessity for music, but we retain vestiges in nightclub and mood music to promote pairing, national folk music to demarcate kingdoms, and martial music to inspire soldiers.

Sound is a remarkably efficient communication channel. It can convey more information than cumbersome visual or olfactory communication. Sound bends around corners and allows an animal to broadcast its territorial claims without traversing its entire domain. So it is not surprising that auditory communication has played a major role in natural selection among animals.

The frequency range of this communication is enormous. While humans are able to perceive sounds between 20 and 20,000 hertz (cycles per second), the range of animal perception is much broader. The well-known ultrasonic sounds of dolphins and bats used for echo-location can approach 200,000 hertz. Some birds, such as pigeons, and even elephants, broadcast low-frequency "infrasonic" sounds that are sometimes less than one hertz.

A combination of genetic and cultural evolution explains the performance of animals that have specialized calls akin to human songs. By removing birds from their nest when very young, biologists have demonstrated that auditory isolation left them able to produce only raucous sounds. During the first months of life, if some male songbirds are deprived of the opportunity to hear other birds sing, they later sing poorly, even when exposed to good singers. But those listening to *recorded* birdsongs played for them as early as ten days after hatching have sung proficiently. An hypothesis called the "auditory template theory of song learning" is that songbirds have some innate capacity to develop at least some of the normal features of song as heard in the wild. A bird hearing normal song modifies its internal "template" with learned information to match the local dialect of calls. This parallels human learning in which innate linguistic syntax serves as the foundation for a particular language or dialect. Crudely speaking, "the same hardware, but different software."

Remarkably, some researchers have found that birds can memorize a song when young and then later break it into "syllables" that they rearrange into new songs. Using these phonetic units, a bird may conform to local song fashions, yet each male territorial bird can establish his own musical trademark. And birds have the unusual ability to sing two songs at once, because of separate vibrating membranes that are akin to human vocal cords, two of them lying at the bifurcation of avian windpipes. Michael Bright writes in his compendious *Animal Language*, "American thrushes are fine examples of birds that can sing two songs simultaneously, but with a harmonious relationship between each rendition, producing some extraordinary tonal qualities which, some researchers feel, are not to be matched anywhere else in birdsong."

Much as bird populations have their geographically distributed song dialects, undersea denizens—whales—sing songs with regional flavors. The songs of humpback whales are long, repetitive vocalizations that last from five to more than thirty minutes, with a song being repeated many times over periods as long as a day. Humpback whales in a certain locale sing the same song. For example, whales from the Pacific Ocean sing different songs from those in the Atlantic Ocean, but the basic song structure is the same for both populations. Zoologist Roger Payne and his former wife Katherine, a musicologist turned zoologist, have documented the cultural evolution of humpback whale songs. The songs change continually, components of a populations's song rapidly evolving to new forms every few months. It is clear that this is cultural evolution in the domain of music—perhaps not so surprising for beings with the largest brains of any animals that have ever lived.

One of the primary uses for animal songs and sounds is species identification. This utility is pronounced for birds, but its quieter appearance in the insect world is more surprising. For example, more than 2,000 species of fruit flies have been identified, and among these, hundreds of distinct "songs" created by rapidly vibrating wings can be discerned. Researchers using highly sensitive sound-measuring techniques have found that fruit flies use these songs, heard a few centimeters away, to identify and then mate with other members of their species.

Music seems to evolve among the lower animals through genetic and cultural transmission. Musical communication that gives some individuals a survival advantage promotes the dominance of that biological strain. The whole process may have begun hundreds of millions of years ago when primitive movements of animal bodies, not designed to produce sound, were fruitfully noisy, leading to advantageous communication between creatures. Michael Bright speculates that the first meaningful sounds may have come from primitive trilobites, 500-million-year-old relatives of the horseshoe crab: "We shall probably never know, although someone, some day may find a trilobite with structures that can only be interpreted as organs for producing sound."

Though the identity of that first primitive note may never be known, we can confidently surmise that music has had a continuing influence on human evolution. It is hard not to imagine that the peoples of prehistory got some of their musical inspiration from birds and other wildlife. Some have speculated that human consciousness and language blossomed simultaneously during the last 100,000 years, impelled by still disputed social forces. The role played by human mimicry of animal sounds may never be known, but interspecies copying is observed even today and might have been part of the evolution of human language.

In historical times music has influenced human culture in ways that are undeniably mystical. Twenty-five hundred years ago the Greek philosopher-mathematician Pythagoras and his colleagues were dimly aware of the relation between frequency of vibration and audible pitch. But more than two millennia would pass before Galileo and the French mathematician Marin Mersenne put the musical vibration of strings on a firm mathematical foundation. The Greek philosophers were impressed by the mystical whole-number ratios in the lengths of vibrating strings that produced pleasing harmonies. They connected these harmonies with the perceived spacing of imaginary celestial spheres on which the seven classical planets (including the Sun and the Moon) were supposed to revolve. To the Greeks, it was no coincidence that the seven notes of their scale were in melodic synchrony with the seven known "planets."

In Guy Murchie's mystical vision of science, *Music of the Spheres*, he quotes Hippolytos (c. 400 B.C.): "Pythagoras maintained that the universe sings and is constructed in accordance with harmony; and he was the first to reduce the motions of the seven heavenly bodies to music and song." Thus was born the idea of "music of the spheres." This mystical speculation induced by music helped send Greek science down a path that we are still following.

In 1596 astronomer Johannes Kepler published *Mysterium Cosmographicum* ("Mystery of the Cosmos"), his failed attempt to relate the spacing of the planetary spheres to the geometry of the mysterious five regular Platonic solids (the cube, tetrahedron, octahedron, dodecahedron, and icosahedron)—the only solid shapes possible with identical faces (the solid must be "convex"—i.e., without bowing-in). Through his frustration with this pursuit, Kepler was driven to discover the true principles behind planetary orbits. The three Keplerian laws soon helped Isaac Newton develop his theory of universal gravitation that was so seminal in the development of western science. Kepler, not satisfied with his bare mathematical laws, later published a musical interpretation of each planet's "harmony," *Harmonica Mundi* ("Harmony of the World"). Recapitulating Pythagoras, he formulated a mystical connection between the vibrations of strings and the velocities of planets.

Astronomy in particular, but the sciences in general, have attracted people who were also musically inclined. The great eighteenth-century English astronomer William Herschel earned his living as a notable musician and composer before he found a patron for his astronomical observations. Albert Einstein was as fond of his violin as of theorizing. Musical scientists are legion, and some have suggested that musical patterns subconsciously attract scientifically or mathematically minded people with the same emotional magnetism as do regularities in nature.

Some scientists have gone beyond emotional or mystical

tinkering with music and have tried to fathom underlying patterns of its aesthetic appeal. The problem in discovering these is that much good music seems precariously balanced between confusion and regularity. Composer Lejarin Hiller, Jr., says, "Music is a compromise between monotony and chaos." Guy Murchie writes, "A composer with an eye to physical theory need only make the entropy [disorder] of his music low enough to give it some recognizable pattern yet at the same time high enough for an element of surprise and individuality, and he may be well on his way toward a judicious compromise between the Scylla of wanton discord and the Charybdis of dull monotony!"

Taking up this tightrope challenge, physicist Richard Voss has discovered the manner in which good music may reflect statistical properties of the natural world. Voss has built his computer-generated music around the ideas of Benoit Mandelbrot, whose theories of natural geometry seem to apply as much to music as to the visual arts. With his "fractal geometry of nature," Mandelbrot found order in seeming disorder—mathematical regularities in things as disparate as the convolution of clouds and the jaggedness of mountain ranges.* Voss now suggests that computer-generated music with "fractal" statistics is much more pleasing than earlier computerized music without that natural flair. Martin Gardner, describing this work in *Scientific American*, says, "One can create 'mountain music' by photographing a mountain range and translating its fluctuating heights to tones that fluctuate in time. If we view nature statistically, frozen in time, we can find thousands of natural curves that can be used in this way to produce stochastic [statistical] music." One geneticist has even used the molecular pattern in DNA to generate music. Good music mixes order and surprise. Gardner says that Voss has given

---

*Fractals are a family of intricate shapes that tend to have the same "degree of irregularity" no matter at what magnification or scale they are observed.

mathematical measure to this well-known mixture: "How could it be otherwise? Surprise would not be surprise if there were not sufficient order for us to anticipate what is likely to come next. If we guess too accurately, say in listening to a tune that is no more than walking up and down the keyboard in one-step intervals, there is no surprise at all. Good music, like a person's life or the pageant of history, is a wondrous mixture of expectation and unanticipated turns."

So music seems to be encoded in nature in myriad ways. And since musical elements are omnipresent in the terrestrial biosphere, it is not hard to believe that music, like life, may similarly pervade the wider cosmos. The music of vibrating matter gave rise to life, which now contemplates its images and destiny in the skies. Silent symphonies may secretly surround us, their performance locked selfishly in the bosom of distant worlds. Some may be forever silenced, as civilizations and worlds fade and die. But other symphonies may endure in the barren reaches of space, deliberately dispatched electromagnetic vibrations mingling with the echos of the Big Bang, waiting to caress a willing or indifferent planet.

Dare we doubt that capturing music sent from the spheres may have the greatest import at this moment in cosmic evolution? It is not the silent "absence of evidence" we crave, rather the evidence of endurance. We should seek, not fear the siren song. For it may be that life's "destiny," if we may be forgiven that mystical term, is as much bound with the course of other civilizations as it is with the fate of the universe.

We have sent our own music inadvertently and deliberately out into the forgetful universe—on pulsating electromagnetic ripples and borne by spacecraft. Two copper discs slowly, silently traverse space—on them the sounds of animals and the songs of people, introduced by the haunting music of the planets, Kepler's *Harmonica Mundi*. Silent symphonies—music beyond the spheres.

DESTINY

13
THE FATE OF THE UNIVERSE

Some say the world will end in fire,
Some say in ice.

—ROBERT FROST, "Fire and Ice"

Perhaps a species that has accumulated ten tons of explosive per capita *has already demonstrated its biological unfitness beyond any further question.*

—ARTHUR C. CLARKE, 1983

I believe that man will not merely endure: he will prevail.

—WILLIAM FAULKNER, *December 1950, On receiving the Nobel Prize*

When quacks with pills political would
dope us
When politics absorbs the livelong day,
I like to think about the star Canopus,
So far, so far away.

—BERT LESTON TAYLOR, "Canopus"

It may seem absurd to ponder whether intelligent beings—as we usually pride ourselves—have any significant role to play in the destiny of the cosmos. Moreover, for entities not accustomed to worrying about events much further removed than a few years or at most a decade, speculating on the happy or sorry fate of the universe billions of years from now may seem an unlikely exercise, irrespective of whether we have any power to alter the larger course of the cosmos.

Yet with spectacular leaps of imagination, cosmologists have already considered how *we* might someday transform the universe, and perhaps even save it and our own precious future hides from oblivion. In fact, cosmologically in vogue is the study of the "fate of the universe," even how it may be affected by the strivings of intelligent life. The ultimate questions of this cosmic eschatology are: does life have an important role to play on the grandest scales of time and space, and, if so, does it have a prayer of a chance to preserve itself in the universe's fiery collapse or eternal icy expansion?

What is it in the pageant of cosmic evolution that we would have any interest, much less the power, to change? Are we not like hapless beetles riding a piece of drifting flotsam that is destined to plummet down a waterfall? Aren't we merely along for the ride as the universe expands and plunges toward an unknown senescence? Physicist Freeman Dyson, among the first to tackle this question, answered boldly in 1979, "It is impossible to calculate in detail the long range future of the universe without including the effects of life and intelligence." He immediately added the thorny caveat: "If we are to examine how intelligent life may be able to guide the physical development of the universe for its own purposes, we cannot altogether avoid considering what the values and purposes of intelligent life may be." Morality and "purpose" blended unexpectedly and irreversibly with physical law!

Consider the vast energies that religious thinkers and philosophers have lavished over the ages on the survival of an intangible "soul" beyond an individual life. In that light, is it so surprising that cosmologists interest themselves in what lies far beyond the present incarnation of the universe? Though their methods and assumptions are indisputably different, a greater philosophical kinship connects cosmologists and those who ponder the fate of evanescent souls than either group may be willing to admit. They both embody life's age-old quest to transcend the ordinary boundaries of everyday existence.

It is impossible to peer trillions of years into the future without first looking back at the few billion years that have transpired to chart the first halting footsteps of life, and then looking forward a few short aeons into the future. Life, at least as we know it, has in one sense already begun to transform the cosmos, and yet quickened matter arose barely four billion years ago, according to best evidence. The universe, so cosmologists believe, is about 15 billion years old. Earth's biosphere, that thin shell of air, soil, and ocean, already appears to be fully controlled by life. The planet possesses an oxygenous atmosphere only by the exhalations and deaths of countless plants and microorganisms. Life has reworked the surface of the world, and appears to be acting on a global scale as a single giant organism—Gaia, regulating internal temperatures and chemical concentrations much as do individual beings, homeostasis with a planetary scope. So on one world we have a definite analogy to what might happen someday on an interstellar or even an intergalactic scale: life altering its environment far beyond the locale of individuals.

Earth is but one tiny yet precious dust mote of a planet, quite lost in the depths of space. Suited hominids have just stepped off this world and are now testing the waters of the cosmic ocean. Fresh from conquering the air with silver wings in less than a century, they now lay tentative claim to much of the solar system. If human civilization can resolve its passionate fraternal struggles and not self-destruct, it seems probable that virtually nothing will prevent the colonization of the planets, moons, asteroids, comets, and barren space in the Sun's kingdom.

Within several hundred years, perhaps far sooner, people may live on dozens of distant worlds. Some of those orbs may be changed forever by human design. Realistic plans are afoot to "terraform" Mars—to remake it in the image of Earth by gradually changing its climate and atmosphere. Even hellish Venus may submit to human desires for clement worlds. Astro-engineering on the scale of a planet has been discussed for so long that it is almost passé.

Enlarging the scope of solar system re-engineering, Freeman Dyson suggested years ago that a civilization might actually disassemble the planets of its childhood and construct a spherical network of orbiting structures completely enclosing its star. Within this "Dyson sphere," the interplanetary civilization could capture nearly the total energy output of its mother sun. Dyson even proposed that astronomers seek evidence of extraterrestrial civilizations from the infrared radiation that would necessarily be emitted by such stupendous stellar envelopes.

Unfortunately, even long-lived stars like our Sun don't shine benignly forever. Old Sol has been fusing its hydrogen into helium for nearly five billion years, providing energy for earthly life for much of that time. But about five billion years hence, with its hydrogen fuel reserves dwindling, the Sun will enter a so-called red giant phase. Its bloated body will extend beyond the orbit of Earth and possibly make conditions inhospitable, even within the Dyson sphere that may by that time enclose the Sun. This may be the high-water mark for life and intelligence remaining in the solar system. If by then it has not already done so, life may have to abandon its once nurturing and now malevolent star.

It is almost impossible to project with confidence what our civilization might be like 1,000 years hence, much less billions of years in the future, now that we have unleashed the Furies of technology. Contemporary civilization may resemble more closely the ancient Egyptians or Greeks than society in the year 3000. Judged by contemporary newspaper headlines, we aren't that far out of reach of the Middle Ages, a mere eye twitch ago in cosmic history. Yet the pace of evolution is indisputably quickening here on Earth. Driven by technology, history is accelerating.

It is distinctly possible that we will eventually decide to

engineer biological changes in humans deliberately and achieve physical symbiosis with compact forms of electronic artificial intelligence. These trends are already under way—increasing reliance on computers and biotechnology, perhaps one day culminating in a form of biological immortality or the abandonment of biology altogether. Perhaps our psyches will eventually comfortably inhabit nonbiological "machines." If the thought brings discomfort, try to imagine these progeny as loving "children" who are as much (and more) products of ourselves as our own genetically endowed offspring.

Biological and cultural changes are arguably more difficult to foresee than the physical expansion of civilization into interstellar space. The distances between individual stars are staggering and humbling—typically 250,000 times the 150-million-kilometer distance from Earth to the Sun. But there are many ways we already know to cross this gulf, even constrained by the seemingly impassible barrier of the speed of light. We know that engineers and physicists have worked for years to plan the technology of interstellar travel. Probably in the next century, and almost certainly within 200 years, robotic probes will make journeys to the nearest stars that may take only decades, rather than the millennia our present "slow boats" would consume. Peopled missions to the stars will be far more costly, but when sufficient motivation has developed (perhaps combined with extended lifespan, hibernation, etc.), human beings will venture to worlds around the stars.

This leads to a vision of our own civilization expanding in a wave of colonization, each stellar outpost giving rise to more remote ones, and by geometric progression eventually encompassing billions of solar systems in the Milky Way galaxy. The possible transformation from an interplanetary to an interstellar civilization has been calculated to occur in far less than a billion years—a small fraction of the age of the universe. This has led to the alleged mystery and heated debates about why we don't see extraterrestrials crawling all over the place—the Fermi Paradox. (Of course, some would suggest that they *are* here, since presumably they too could ignite waves of interstellar colonization.)

Suffice it to say that there are as many good explanations for the lack of abundant evidence for extraterrestrials as there are stars in the sky—not the least of which may be that aliens may be far more crafty and careful not to disrupt the local biota than we imagine: the renowned Zoo Hypothesis. And the explanation for this apparent nonpresence does *not* have to be a pernicious anti-Copernican doctrine championed by astrophysicists Frank Tipler and John Barrow in recent years: the unabashed claim using intricately tortured logic that we are likely to be the only civilization in the universe.

Once over the conceptual hurdle that the distances between stars are insurmountable barriers to the physical extension of life throughout the cosmos, we can imagine all manner of galactic empires and intermingling of independently evolved life-forms. Perhaps we are simply not aware that we lie in the midst of one such galactic communications net that traffics in artifacts as well as invisible electromagnetic messages. When we gaze at the heavens, we don't see any remarkable patterns of stars that spell out "Eat at Joe's," or the slightest sign that extraterrestrials are zipping around remaking the cosmos for their own needs. This is not to deny that such monumental reorganization of the universe by life may yet happen. Dreaming wildly, we could imagine subtle artificial changes already taking place, ones we simply haven't recognized—stars disappearing right above our noses, so to speak, digested by unimaginable machines. Or it may still be the springtime of the galaxy, and Earth may be among only a handful of civilizations just beginning to realize their potential to change the course of cosmic history.

What would be the course of the universe if we and other sentient life, acting in a spirit of cosmic environmentalism, did nothing to upset natural balances? The universe would continue for at least tens of billions of years its spectacular expansion that began about 15 billion years ago. Almost all cosmologists accept the overwhelming evidence that the universe—all of space, time, and matter—began to expand explosively from an inconceivaby small region. "Before" this Big Bang, we are faced with literally nothing.

On the grandest scale, galaxies are fleeing from one another, but it is really the continuum of time and space that is stretching bubble-like, carrying with it the galactic islands of stars. The central question of modern cosmology is whether the expansion will go on forever in what is called an "open" universe, or perhaps will turn around, becoming a general contraction and ultimate collapse—a "closed" universe. The answer depends on how much gravitating matter the cosmos contains to retard the expansion. But we have seen that there are compelling reasons to believe that the universe is so close to the borderline between open and closed that it may be impossible to tell which it is. Most of the matter in the universe—possibly 90 to 99 percent of it—may not be visible, ordinary matter at all, but the ethereal pervasive background of peculiar dark matter. For all practical purposes (and whoever accused cosmology of being practical?), the universe of ordinary and dark matter is likely to expand forever.

Whatever the scenario, physicists have already calculated the ominous events from here to eternity, assuming we and "others" do nothing to alter them. The numbers of years to these remote eras are so unimaginable that one can only begin to visualize them as *multiples* of the present age of the universe, 15 billion years becoming one "unit" in this chronology. By 10,000 units into the future (10,000 times the current age of the universe), all the stars will have spent their nuclear fuels and will be mere shadows of their once glorious selves—pallid glowing cinders. By 10 million units, all planets will have been torn from stars by chance disruptive close encounters with other stars. By 100 million units, near misses between stars will have kicked or "evaporated" all of them out of galaxies, leaving behind a voracious black hole in each galaxy to devour the little remaining matter.

By 100 billion billion units, if current theories of matter are correct, all the protons in all the stars will have decayed to a mixture of radiation and the subnuclear particles electrons and positrons. In the final act at 10^{90} units, the universe's remaining black holes may have all evaporated and the cosmos may then consist almost entirely of radiation.

Many of these events will occur whether the universe expands indefinitely or collapses on itself. If enough gravitating matter does exist to turn the tide of expansion, at some point space-time will begin to contract and the universe will again heat up to an enormous temperature before zapping out of existence like a burst soap bubble. Alas, there may be a chance to replay the cosmic drama of infinite joy and bottomless woe after such a collapse. No cosmologist is certain, but some believe that the contracting cosmos could "bounce" and reemerge many times in succession, universe after universe popping into existence. Small consolation to our poor world—all will be contracted to oblivion. Even the perennially optimistic and inventive Freeman Dyson is stumped: "Is it conceivable that by intelligent intervention, converting matter into radiation and causing energy to flow purposefully on a cosmic scale, we could break open a closed universe and change the topology of space-time so that only a part of it would collapse and another part of it would expand forever? I don't know the answer to that question."

If the closed universe engenders feelings of claustrophobia and depression, the open universe delights the imaginers of the distant future. Like the prophetic words of ancient seers, the cosmologists' equations foresee an era when intelligent "life" has long since transformed itself to unimaginable ethereal forms, perhaps gossamer collections of nuclear particles tenuously linked together over vast distances. The object of such evanescent life: a fantastic struggle to preserve memories and consciousness in a cosmos growing colder by the aeon as supplies of usable energy dwindle. A hint of this destiny appeared in a 1929 essay, "The World, the Flesh, and the Devil," by British crystallographer J. D. Bernal, who wrote, "Consciousness itself may vanish in a humanity that has become completely etherialized, losing its close-knit organism, becoming masses of atoms in space communicating by radiation, and ultimately perhaps resolving itself entirely into light." Isaac Asimov portrayed an even more tenuous existence of consciousness at the end of time in his captivating science fiction story, "The Last Question."

Physicists do not, of course, presume to know exactly how ghostly beings would be constructed. But they employ the esoteric equations of thermodynamics, information theory, and general relativity to establish whether information—the essence of life—could theoretically exist at all under scarcely imaginable physical conditions.

Nineteenth-century physicists were obsessed with the idea that the universe would ultimately suffer a dreaded "Heat Death" as all matter gradually approached an equilibrium temperature and maximum disorder or entropy ensued. With everything at the same temperature, no useful energy would be available to do work—no thoughts could exist in such a world. This was thought to be the inexorable toll of the Second Law of Thermodynamics. But the expanding universe and modern gravitational theory have now exorcised the Heat Death demon in an unexpected way. Physicists no longer believe that the cosmos will "run down" in exactly the manner that their forebears imagined. They say that though entropy (disorder) will constantly increase, its level steadily falls behind its maximum possible value as the universe expands, thus keeping open the theoretical possibility for organization to continue.

Physicist Steven Frautschi, who extended Freeman Dyson's

work, wrote in 1982, "We have thus come to a conclusion which stands the 19th-century model on its head. Far from approaching equilibrium, the expanding universe falls further and further behind achieving equilibrium. This gives ample scope for interesting nonequilibrium structures (e.g., life) to develop out of the initial chaos, as has occurred in nature."

Alas, another form of cosmic abyss arises phoenix-like from the ashes of the Heat Death devil. In essence, the new threat is that the density of disorder declines in the expanding universe, but that it gradually approaches a low but changeless limit. Yet the growth and development of life necessitates increases in entropy, at least in local regions. You can't have growth without entropy change. Instead of everything just "melting" into a soup of consant temperature as in the nineteenth-century picture, life processes of the far distant future would be perpetually in danger of freezing to a grinding halt—life petrified in a changeless state. No energy would be available in isolated regions to run any kind of life.

Frautschi suggests an out: "A sufficiently resourceful intelligence inhabiting a critical universe learns how to move black holes, bringing them together from increasingly widely separated locations and merging them to increase entropy. The maintenance of life involves a compromise: the entropy must increase, but not so rapidly as to reach a maximum."

Dyson theorized that ethereal life could live eternally by husbanding energy resources and hibernating for extended periods between ages of active contemplation—one of the major functions of life in that penultimate state. He concluded sanguinely that "No matter how far we go into the future, there will always be new things happening, new information coming in, new worlds to explore, a constantly expanding domain of life, consciousness, and memory. I have found a universe growing without limit in richness and complexity, a universe of life surviving forever and making itself known to its neighbors across unimaginable gulfs of space and time." Frautschi is less certain of the viability of Dyson's projections, but concludes that "If radiant energy production continues without limit, there remains hope that life capable of using it forever can be created." We can imagine our ethereal descendants looking back from the remote future on their origins in contemporary civilizations—having the same sense of awe with which we ponder our own distant beginnings among jousting molecules.

Cosmologists occasionally make strange bedfellows. Frank Tipler and John Barrow, while steadfastly maintaining that Earth almost certainly harbors the universe's only "intelligent" civilization, nonetheless ally themselves with Dyson and Frautschi. Tipler and Barrow posit what they call a "final anthropic principle" or FAP, which states that life, once begun, will be impossible to destroy. They say if it were otherwise, the very fabric of the universe would collapse, because there would be no observers to keep the universe observed and going! The bizarre dictates of quantum mechanics lurk around every corner. In a manner of speaking, "The Moon isn't there if there's no one to look at it."

In their book *The Anthropic Cosmological Principle*, packed with formulae and philosophy, Tipler and Barrow end with a conclusion of truly Messianic proportions: ". . . life will have gained control of *all* matter and forces not only in a single universe, but in all universes whose existence is logically possible; life will have spread into *all* spatial regions in all universes which could logically exist, and would have stored an infinite amount of information, including *all* bits of knowledge which is logically possible to know. And this is the end." But not quite, because in a footnote they add that a modern theologian might agree that life at this point would then be "omnipotent, omnipresent, and omniscient." We biophiles can say to that a hearty, "Amen."

Cosmologist Paul Davies suggests that as the expansion of the universe proceeds, eventually nonuniformities in different directions of expansion will generate a heat-producing "shear."

The unequal heating of different parts of the universe could theoretically provide an inexhaustible supply of energy for the preservation of life's ghostly descendants. In Davies' words, "Maybe the universe is immortal after all."

Eternal life taking over this universe and becoming one with it may sound hopelessly Utopian, grandiose, or simply crazy in the extreme. Yet faced with a cosmos that seems (by most accounts) to have every "intention" of expanding forever, we are obliged to wonder how it will all unfold. The idea that intelligent life could play a major role in shaping the cosmos may be either a wild delusion or one of the most sobering thoughts imaginable. The imponderable ages over which this spectacular evolution would occur tortures our conventional concepts of time. With smug self-satisfaction, while being virtually certain that it can't be so, we sometimes imagine ourselves to be at the pinnacle of existence, 15 billion years removed from the earliest moment. If that time is as one year, then all recorded history occurs in the last seconds before midnight on New Year's Eve, a sobering enough comparison. Profoundly deflating our boundless egos, it now seems that the first 15 billion years may be nothing more than the first fleeting instant of what is to come. On this scale, life has exploded onto the scene almost concurrently with the birth of the cosmos. We have a long, long way to go.

It is impossible to avoid the impression at this juncture that some of the concerns of traditional theology and forward-looking cosmology appear hopelessly entangled. A peculiar theo-cosmology, in fact, seems to be on the rise—a cosmic philosophy without rigid doctrines that treat human civilization as though it were the only one in the universe. And Freeman Dyson set forth its fundamental tenet: "If our analysis of the long-range future leads us to raise questions related to the ultimate meaning and purpose of life, then let us examine these questions boldly and without embarrassment." Open questions, indeed, for an open universe.

14

THE GRAND ILLUSION CALLED TIME

Of all the obstacles to a thoroughly penetrating account of existence, none looms up more dismayingly than "time." Explain time? Not without explaining existence. Explain existence? Not without explaining time. To uncover the deep and hidden connection between time and existence, to close on itself our quartet of questions, is a task for the future.

—JOHN ARCHIBALD WHEELER, 1986

Only a cosmic jester could perpetrate eternity and infinity. . . .

If eternity is silliness, then infinity of space is sheer madness.

—EDWARD HARRISON, *Masks of the Universe,* 1985

The key to comprehending space-time is the obvious (to me) fact that space is the relationship between things and other things while time is the relationship between things and themselves.

—GUY MURCHIE, *The Seven Mysteries of Life,* 1978

We speak of cosmic evolution and ponder the fate of the universe, implying the passage of time. Our "cosmic time" is clocked by the geometric expansion of the cosmos. Humble beings locked in the matrix of time and space, we too easily take time for granted. Instead, we should pay homage to the enduring mystery of omnipotent time.

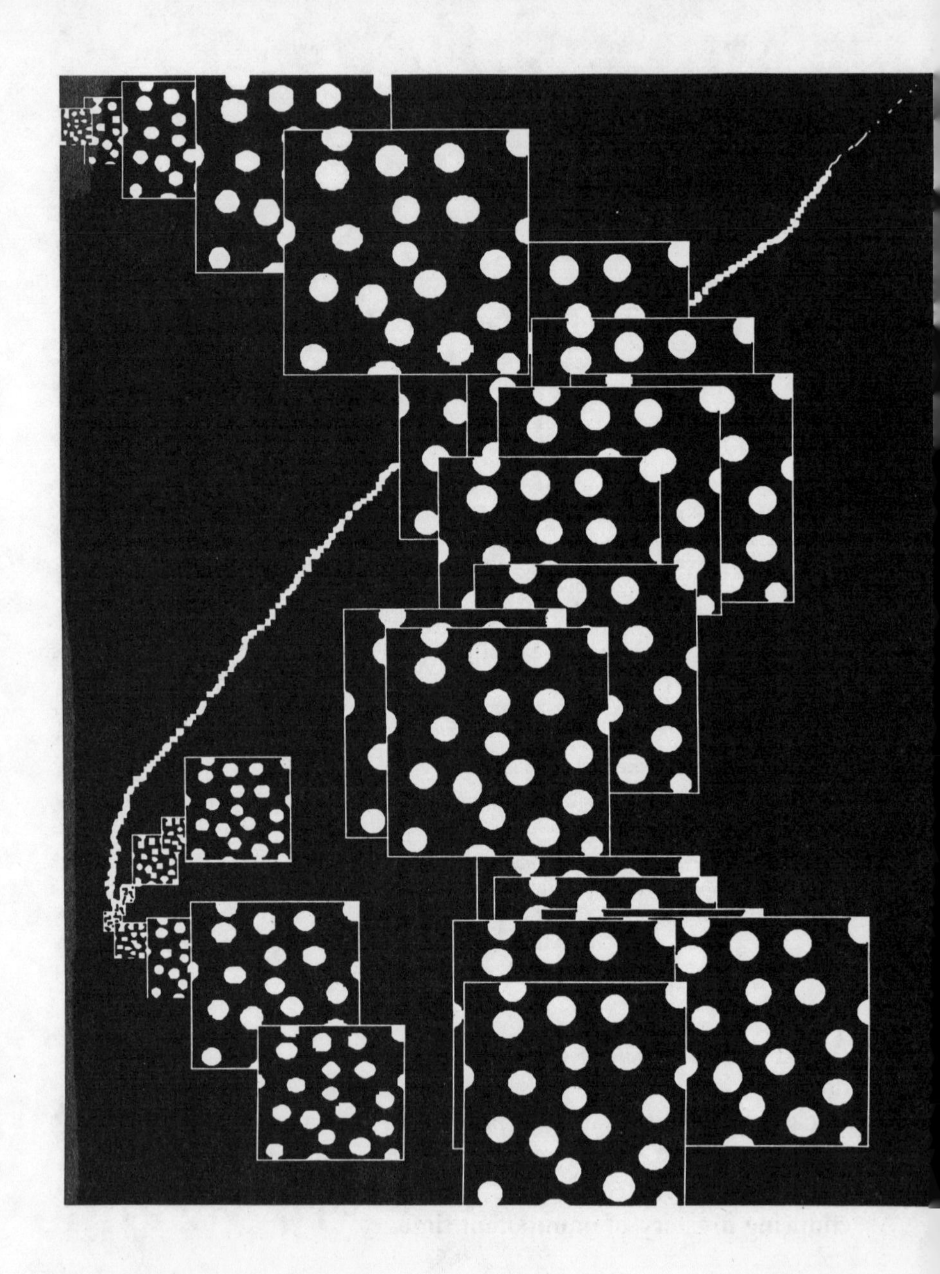

We are all time travelers. We plunge toward an unknown future, fleeing from an ever dimming past, barely able to savor the present. The concept of time is so entwined in the grammar and meaning of language that it is difficult for us travelers to step back and see time for what it really is. According to the best scientific theories, our perception of time is an illusion—perhaps the grandest one in nature. Our seemingly rock-solid notions of "now," "past," and "future" crumble beneath mountains of scientific evidence that time is not what it seems.

On the level of subjective time, we already knew that something was amiss. Boring tasks seem to consume hours, while stimulating ones rush by in what feels like minutes. A child's summer vacation seems to him an eternity, while for an adult it feels like the short span of three paychecks. Experimental subjects removed from all clocks and external evidence of daily cycles soon begin to underestimate how much clock time has transpired. We can chalk up these distortions to the inscrutable workings of human consciousness and then try to rest on the laurels of our finest atomic clocks. Surely these modern wonders, slipping barely one second in 3,000 years, tell time's story straight—or do they?

All time reckoning depends on the unfolding of a pattern of spatial happenings. From the grains of sand falling through the hourglass throat to the furious regular beating of cesium atoms in an atomic clock, we count time by the occurrence of events in space. Is there a disembodied time—time without space? If space were devoid of all objects would there be time? The mind recoils from this concept of a "timeless space" and instinctively imagines a metaphorical "stream of time" flowing even in that empty space. What about time's *direction* in an empty space? Past and future would be hopelessly confused in such a realm. Boggling the mind a bit more, try to imagine the absence of space itself—a favorite pastime of cosmologists apprehending the beginning of the universe. Could there be time without space? Hardly.

As mysterious as the "flow of time" are its beginning and end, if either there be. Can anyone point to the source and sink of the river of time? Religious thinkers may have spent almost as much effort searching for an answer as they have pondering the whereabouts of souls before and after the great levelers—birth and death. Though it is not the province of science to probe the mysteries of intangible soul, science does have much to say about the beginning and end of time. Modern science believes that time and space did have an instantaneous beginning about 15 billion years ago, give or take a few billion years of cosmic time. We have seen that the jury is not yet in on whether time and space will come to an inglorious end, but it is swaying toward infinite continuation. A smaller scientific faction favors a rebounding time—the universe of space and time restarting after collapse in the distant future, a universe of many if not infinite incarnations.

From simple musings of this paradoxical sort, the idea emerges that time and space must be intimately connected. This was indeed the view that Albert Einstein proposed in 1905 and 1916 in his revolutionary theories of Special and General Relativity. Apart from its enormously successful predictions of all kinds of phenomena, relativity has brought about a true revolution in thinking about time. "Common sense" requires that what is "now" for me is the same for everyone else, that my past and future are the same for the entire universe. For everyday commerce, these common sense notions suffice and are correct to a high degree, but according to relativity they are fundamentally flawed and totally inadequate under more exotic conditions. There is no universal "now," "past," or "future." Each frame of reference (all objects not moving with respect to one another define a frame of reference) has its own local time, and time appears to run at different rates depending on how fast one frame is moving with respect to another. Time is a very private matter. Moreover, in the neighborhood of a gravitating mass, time appears to run slower than in space far away from the mass.

So the first illusion of time is its absolute, steady flow throughout the universe. The lovely metaphor of the "stream of time" is nothing but a cruel hoax. There is no more dramatic proof of the illusion than the so-called twin paradox. It is really not a paradox at all with the proper understanding of relativity, but the scenario certainly affronts common sense. We have two twins on Earth, one of whom blasts off in a super-rocket (far beyond current technology, of course) and accelerates to 99 percent of light speed. The astronaut twin drifts for some years, then decelerates, turns around, and again rockets toward Earth near light speed. She politely decelerates near Earth to soft-land near her twin sister. Upon greeting one another, the sisters realize that only five years have transpired for the astronaut—she is biologically only five years older, consistent with the rocket's clock. Yet the Earth-bound sister has aged thirty years by her reckoning, biologically and clock-wise.

Strange as this circumstance appears, and as far away we seem from enacting it with people, the physical and mathematical basis of the twin scenario has been verified many times over in laboratory experiments with accelerated nuclear particles traveling near light speed. It is often remarked by some who freely accept relativity's time-dilation effect (the slowing of time for relatively moving frames), that there is still a paradox, because during the flight each of the twins reckons that the other's time has slowed down. Unfortunately they neglect the asymmetry of the twins—one is privileged to go through many stages of acceleration and deceleration. Exact calculation of the astronaut twin's motion proves that there is an age discrepancy on reunion. And the closer the super-rocket gets to the speed of light, the more dramatic the discrepancy in age. In principle, it would be possible for the astronaut twin to return to an Earth that had aged millions of years, while still awaiting her promotion in the now defunct Astronaut Corps.

With vast reserves of energy to burn, as well as a few years of your own time, at least one way is available to travel to someone else's future. The lesson of the twins is not this delightful if extravagant application, but proof beyond doubt that time is not what it appears to be. How had this illusion escaped physics for so long? One of the clearest expositors of relativity, Albert Einstein, wrote in 1936, ". . . that the meaning of physical time is a priori clear—this illusion had its origin in the fact that in our everyday experience we can neglect the time of propagation of light. We are accustomed on this account to fail to differentiate between 'simultaneously seen' and 'simultaneously happening'; and, as a result, the difference between time and local time is blurred."

To further clarify the point, take the often quoted example of a fast-moving freight train with an observer at its geometrical center. Just as the train-borne observer passes another observer at a rail crossing, lightning bolts strike both the front and the back of the train. If the ground observer sees the light from each bolt at the same time, he may be sure that they happened at the same time in his frame, because the distance light travels from both the front and back of the train is the same for him. But the observer riding on the train's midsection sees the forward bolt happen first, since while the light is traveling toward him, the train has moved slightly forward and the forward bolt's light gets to him before the rear bolt (which must also travel slightly farther). Here we have nonsimultaneity of allegedly simultaneous events (to the ground observer) staring us right in the face. Both observers have a correct view of what happened; each frame simply has its own local time. This all depends on the amazing fact that the speed of light (in a vacuum) is constant for all observers—coming from or going toward a source of light!

According to relativity, time must be viewed as another dimension added to the three obvious dimensions of space. In this so-called four-dimensional space-time continuum, events play themselves out. Even more precisely, each point in the continuum *is* an event. Our lives are very timelike lines wending through the four dimensions—we hardly move at all through space. Spacelike lines would be traced by objects moving faster than light speed. But nothing can exceed light speed once it is already slower than light. Some physicists have speculated about imaginary-mass tachyon particles that might always travel faster than light, requiring infinite energy to slow them down below light speed. With these hypothetical particles, one could send signals to one's past, assuming there were any way to generate and receive tachyon beams. For the moment—both mine and yours—tachyons are simply ephemeral and at this point a bit tacky science fiction.

Even though to understand relativity conveniently physicists must rely on four-dimensional space-time, the time dimension is not equivalent to the space dimensions. So don't be unhappy if you don't feel like jumping up into time. For one, in order for the axis of time to have the same units as space (e.g. kilometers), time must be multiplied by the velocity of light (299,792.5 kilometers per second).

If one is intent on being a time traveler to the future, there are two ways to go, other than super-rocketry. One is simply to hibernate. Better still, suspend all body processes indefinitely by careful deep-freezing and hope that someone will be able and willing to wake you up in the future. Since hibernation and suspended animation have not been developed, that route will have to wait. The other way is to put to advantage the slowing down of time near a source of intense gravity, such as the black hole from a collapsed star. If one could avoid being torn apart by "tidal" forces near the event horizon of a black hole (the surface from which you can't escape intact), then one could sojourn for a while and come back (if you have powerful enough rockets) much younger than the civilization left back on your home planet. In fact, physicists believe that at the event horizon of a black hole, time comes to a complete stop.

Another strange aspect of time: events always appear to proceed in one direction. We can travel backward or forward in space, but we seem unable to do the same for time. It looks like there is an "arrow of time," and it points to each of our futures, never to the past. This apparent arrow of time is one of time's greatest tricks, and a subject still hotly debated by physicists. The problem is that the laws of physics, including mechanics, electromagnetism, and even twentieth-century quantum mechanics, all work effectively and have meaning with time running backward. In essence, substitution of $-T$ for $+T$ in all the equations of physics describes motions and states of particles that are every bit as valid as forward-running time.

If you view the separate frames of a movie depicting the motion of ideal (frictionless) billiard balls bouncing around a table, there is simply no way to tell how to order the sequence of frames. Running forward or backward, the movie makes as much sense to the laws of physics. Time's arrow is lost. Why, then, are we all born, live out our lives, and die, when at the discrete particle level of every atom in our bodies this whole process could run backward and still make sense to physical law? The answer is only partly glimpsed by modern science. Depending on which physicist or philosopher you consult, it has to do with probabilities, the Second Law of Thermodynamics that ordains the increase in disorder of matter and radiation, and even the expansion of the universe. Israeli physicist Y. Ne'eman, reviewing the matter of time reversal asymmetry, writes, "Theoretical physicists are humbled by the realization that almost 150 years after the promulgation of the Second Law of Thermodynamics, they cannot yet consider the Law and its supposed connection with the arrow of time as a solved problem."

Real processes involving trillions upon trillions of particles always seem to proceed in the direction of increased entropy (molecular disorder). Given a container with clear water on one side of a barrier and dyed water on the other side, removing the barrier allows the dye molecules to mix throughout the entire volume, coloring all the water. One never sees the dye molecules subsequently removing themselves to one corner of the container, leaving the rest clear—though there is a finite but exceedingly small chance that this could happen after a time much greater than the age of the universe elapsed. Though the laws of particle motion don't prohibit the disentanglement of closed, complex systems, the "boundary conditions," or the way particle states of motion are initially set up, make sudden reductions in entropy extremely unlikely.

Some physicists are troubled by blithe acceptance of the Second Law of Thermodynamics as the "explanation" of the arrow of time. Hermann Bondi wrote, "It is somewhat offensive to our thought to suggest that if we know a system in detail then we cannot tell which way time is going, but if we take a blurred view, a statistical view of it, that is to say throw away some information, then we can. . . ." This has prompted some to explain time's arrow on "cosmological" grounds, somehow tying its particular direction to the expansion of the universe, and the large-scale influence of gravity on all matter. Astronomer Fred Hoyle writes, "The thermodynamic arrow of time does not come at all from the physical system itself. It comes from the connection of the system with the outside world. The arrow of time is derived from the largest-scale features of the universe." Taking a different tack, physicist Philip Morrison claims, "Time's arrow is then the necessary consequence of the fact that no physical theory, except perhaps the final one, can describe the whole of the universe." He further argues that "The arrow is real, that is, not subjective, that it is not essentially cosmological, that it arises from an inescapable feature of all physical theory."

There is some evidence that an explanation for time asymmetry may lurk in some very peculiar and unique particle properties at the subatomic level, though many physicists would deny that this is the answer. Since 1964 it has been known that neutral (no charge) *K mesons* or *kaons* exhibit an unusual time asymmetry, defining the direction of time's arrow as do no other particles. Kaons participate in certain decay processes that go in only one direction, never in reverse. Is this a tiny arrow of time pointing the way for the Big Arrow? Probably not, but the unexpected behavior continues to baffle.

In 1945 physicist Richard Feynman developed a clever interpretation of antimatter particles. He showed that they could be viewed as particles of normal matter temporarily moving backward in time. For example, a positively charged electron (a positron) can be thought of as an electron moving backward in time, and this is still consistent with the laws of quantum mechanics. Later a violation of what had been known as "charge-parity (CP) conservation" was discovered—situations involving neutral K mesons in which the combined charge and the left or right "handedness" of particles failed to be maintained in nuclear transformations. In order to restore the assumed physical order, Feynman's time-reversal ideas were brought out and generalized to "CPT" conservation, the conservation of charge, parity, *and* time direction combined.

It is possible, at least theoretically, that entire time-reversed domains of antimatter could exist. Martin Gardner writes in *The Ambidextrous Universe*, ". . . and here we plunge into almost total fantasy—the universe may contain galaxies of antimatter in which all events, micro and macro, are moving backward with respect to our arrow of time. Two galaxies would be time reversed relative to each other in somewhat the same way that they are mirror reversed. In each galaxy, intelligent creatures would be moving forward in *their* time. Each would find events going backward in the *other* galaxy." Physical law may make it impossible for beings in such anti-galaxies to communicate with us. Since their light would be going backward in time, converging on them, we would not even be able to see these anti-galaxies.

Time travel to someone's future is not fraught with the problem of violating causality, while traveling to the past is. We more willingly accept the former and disbelieve the latter. The idea of going into the past and preventing your grandparents from meeting highlights the paradox of backwards time travel. But General Relativity points the way to the possibility of regressive time travel in its description of certain trajectories through a rotating black hole—an object that might form after the gravitational collapse of a massive rotating star. Astrophysicist William J. Kaufmann writes in *Black Holes and Warped Space Time*, "Perhaps, therefore, a rotating hole connects our universe with itself in a multitude of places! These would be different places in space and/or time. In other words, by emerging into one of these 'other universes,' you might actually be re-entering our own universe in the same place but at a different time. This is a time machine!"

Black holes being the sort of unbelievable objects they are, General Relativity or no, there should be a healthy amount of skepticism about travel to the past through a rotating black hole, but the possibility cannot be excluded. Some physicists have sought to eliminate the alleged paradox of travel to the past by taking advantage of one of the interpretations of quantum mechanics—the so-called many worlds view. The many worlds interpretation sees the universe as an infinitely bifurcating structure of parallel universes, each completely out of touch with all the others. At each probabilistic turn (discrete change in a single particle state), the universe forks into a pair of separating universes. Time travel to the past might take an explorer *across* to and back along another branch of this tree of universes. Conditions in that other universe might allow the traveler to fit in without disturbing causality.

There is a sense in which time travel to the past occurs

every day. Light from distant stars and galaxies that is reaching us now set out on journeys from tens to billions of years ago. Simply gazing at the stars is a way of peering back at the past, but it is not a way of being where that local past was happening.

Many ancient cultures considered time to be cyclical on a grand scale of perhaps thousands of years. In this perspective, after a cycle of history passed, the same series of events would occur again and again. In western culture the Judeo-Christian concept of linear time came to replace those antique images of circling time. We became obsessed with time's beginning and time's end. Twenty years ago belief in the steady state expanding universe was scientifically tenable. Time extension to the infinite past of cosmic time—time measured by the expansion of the universe—was a key feature of the steady state cosmology. The muffled birth cry of the Big Bang universe, discovered in the form of cosmic microwave radiation in 1965, ruled out the steady state universe and an infinite past. The universe of space and time did have a beginning, and we have discovered that we are not that far away from it compared to the ages that may remain.

Will time ever end? Yes, if the universe contains sufficient gravitational mass to cause an ultimate collapse. Given enough invisible or "dark" matter, the universe and time may come to an end. But time may also end in an open universe. As the new inflationary cosmology suggests, there may be an amount of gravitating material that just barely allows the expansion to go on indefinitely—slightly shy of the amount for collapse. In the most remote future of that universe, stars and galaxies, planets and people, matter itself will have decayed to an ever more dilute subnuclear soup. In a universe without tangible processes to mark time, time will have neither meaning nor direction. Unless the last vestiges of life have taken steps to forestall its ultimate demise, time, the mysterious grim reaper of life, will itself be dead.

15

Is the World Real?

In the impalpable and seemingly inconsequential entities of the quantum world, one finds the true music and magic of nature.

—Edward Harrison, *Masks of the Universe*, 1985

Human beings are organisms capable of manipulating internal representations of the world by means of concrete operations and can transcend the bounds of their biologically given perception. They can liberate themselves and construct a view of reality that conflicts with intuition, yet gives a true, more encompassing view.

—Max Delbrück, *Mind from Matter?*, 1986

Any view of existence which does not reckon with quantum theory and the elementary quantum phenomenon is medieval. Quantum theory marks the summit of the exact natural science of our day.

—Hermann Weyl, *Philosophy of Mathematics and Natural Science*, 1949

As we climb the mountain of scientific discovery and behold the panorama of cosmic evolution, our conception of reality acquires a quizzical dualism. Fundamentally, we cannot doubt a universe that gives rise to its own self-awareness with such apparently inevitable and intricate process. But far below this lofty seeming height, we glimpse in the quantum river valleys of the microcosm a creeping horror that profoundly shakes our naïve view of reality. The foundations of causality and determinism seem to crumble and the substance of the world dissolves into imaginary waves of quantum

Adapted from drawings by John Tenniel in Lewis Carroll's *Through the Looking Glass: And What Alice Found There*.

mechanical improbability. At rock bottom where rocks themselves become virtually empty space, reality is of a different order than we had ever imagined in our climb toward the summit. Worse, reality may be far different than we ever *can* imagine. Quantum unreality even bubbles up into the macrocosm and betrays our conception of the everyday "I think, therefore I am" universe.

We know that time is out of joint, not what it seems, a preposterous illusion. With that paradox we have made peace. We have even grown comfortably aware that the great chain of being emerged 15 billion years ago through a fantastic elaboration of nothingness. But now that the universe is coasting and purring smoothly, we don't expect it to be contrary or capricious at its roots. Yet it is, and disturbingly so. Despite the best efforts to restore its solidity, it stubbornly resists.

The problems begin with a trouble of atoms. The granularity of the world's microscopic fabric was imagined for thousands of years, first by Greek sages, later perhaps, medieval alchemists, and then classical physicists. Strange though it seems in retrospect, some noted scientists continued to struggle against the atomic picture through the early years of the twentieth century. When evidence for the existence of atoms became overwhelming, it had to be accepted. Yet it is unbearably difficult to believe that these figments of void constitute a solid-seeming world. If we could expand the nucleus of a typical atom to the size of a pinhead, the outer boundaries of its vacuous body would take on the dimensions of a sports stadium. A few invisible geometric points (or vibrating "strings" in latest theories)—electrons—hurtle through the arena and miraculously provide the atom with its blobby geometry, an entity 10^{15} times the volume of the tiny nucleus.

Since the hollowness of atoms was far from apparent, we can perhaps excuse the naïve exuberance of eighteenth-century scientists who imagined a billiard-ball universe. Like billiard balls that bounce and occasionally stick together, hard little atoms were said to be the material foundation of the world. In principle, an omniscient being knowing the exact position, mass, and velocity of every atom at one instant could, by using the newly developed theories of dynamics and gravitation, predict the future course of all matter. Free will was merely an illusion, and determinism ruled the cosmos. Pierre Simon, Marquis de Laplace, boasted:

> We ought then to regard the present state of the universe as the effect of its antecedent state and the cause of the state that is to follow. An intelligence knowing at any given instant of time, all forces acting in nature, as well as the momentary positions of all things of which the universe consists, would be able to comprehend the motions of the largest bodies in the world and those of the smallest atoms in one single formula, provided it were sufficiently powerful to subject all data to analysis: to it nothing would be uncertain, both future and past would be present before its eyes.

The first three decades of the twentieth century eventually shook the complacence of classical physics, and the reverberations of that upheaval are still painfully felt in continuing debates over the meaning of the new theories. First, relativity overturned common sense ideas of time and space. Then the evolution of quantum mechanics during several decades seemed to suggest that an explanation of the world at its most fundamental level would always remain incomplete. Unlike relativity, quantum mechanics didn't spring from the mind of a single theorist. The veil of the microcosm yielded slowly and unexpectedly to probing by many physicists.

A great irony of the new awareness was Einstein's explanation of the photoelectric effect—how light of frequency greater than a definite value can knock electrons from the surface of some metals, while lesser frequencies have no effect. He provided the first theoretical understanding that light, a phenomenon that classically seemed to behave like waves, could simultaneously act as though it were made of discrete energy "quanta"—particles. Einstein never stopped believing that underneath the statistical, probabilistic behavior of quanta lay a deterministic explanation of the world. Yet his unraveling of the photoelectric effect illuminated what has been called the central mystery of quantum mechanics—the duality of waves and particles. The mind recoils from the idea that a wave—a phenomenon distributed widely, even infinitely across space—can act simultaneously like a localized particle. One might sooner believe that an ocean wave was also a grain of sand. By his own hand, Einstein helped launch the descent into quantum unreality.

Without the mathematical formalism of quantum mechanics, some of the most fundamental aspects of nature would be unexplained. Quantum mechanics tells, for example, how an electron can hover around an atomic nucleus without radiating away all its energy and falling in, as classical theory predicts should happen. Yet what does quantum mechanics offer in place? Electrons that jump between discrete states of energy around the nucleus, never landing anywhere in between. Electrons mysteriously "know" at what level to stop when they jump, neither overshooting nor undershooting one iota. And where are they while "in transit"? Simply nowhere.

The essence of quantum unreality is a mathematical description called a *wave function* that fluctuates within time and space. When a particle's wave function is squared (multiplied by itself), an entity smeared across space known as a *probability amplitude* results. This is a mathematical description characterizing the statistical probability of where a particle is likely to be at a particular instant or in what state of energy, spin, etc. A particle may, for example, have a tiny but finite probability of being at a place where, from a classical perspective, it couldn't possibly be. Particles can thus "tunnel" through conventionally impenetrable energy barriers—the equivalent in the macroscopic world of an automobile suddenly lifting off the Earth and appearing mysteriously beyond the Moon.

Quantum mechanics embodies a principle called *complementarity* that introduces a fundamentally irreducible level of uncertainty in the behavior of matter. Complementarity is exemplified by the dualism of particle and wave, but it is much more. Two quantities such as position and momentum form a pair which is in this sense complementary: the uncertainty in knowledge of a particle's position multiplied by the uncertainty in knowledge of its momentum must always be greater than a tiny and precise number called Planck's constant. (Energy and time are likewise complementary.) In principle, one could know the exact position of a particle, but then knowledge of its momentum would be nonexistent. In a very real sense, this uncertainty principle can be thought of as a "conspiracy of nature" to conceal exact knowledge that in theory could provide a deterministic description of the world. But complementarity is even more bizarre: Nature herself possesses no "knowledge" whatsoever of a definite momentum or position of a particle. The uncertainty lies not in the crudeness of our measuring apparatus, but in Nature's fundamental indecision as to where things are or how much momentum they have. But when an observer makes a physical measurement involving quantum mechanical entities, the wave function of the system suddenly "collapses" and at least some aspects become definite up to the limit of uncertainty.

Niels Bohr was a pioneer in a field of luminaries the likes of Heisenberg, Schrödinger, de Broglie, and Planck. Bohr developed the quantum mechanical description of the atom and expounded the profoundly troubling iridescence of quantum mechanics. Casually accepted today by most physicists as a working principle, this so-called Copenhagen interpretation declares that the physical states of particles in the microcosm—their condition of *spin*, momentum, energy, etc.—simply *do not exist* until an observer–experimenter measures them. This has led to much angst, often expressed by philosophical conundrums extended to the macrocosm: "Does a tree really fall in the forest if no one is there to observe it?" or "Is the Moon there when nobody looks?" In short, is there an objective reality beyond observation, and, by implication, beyond consciousness? The "logical positivist" school of philosophy, so congenial to the Copenhagen interpretation of quantum mechanics, would say no, that reality lies only in observations.

Apart from its unresolved philosophical dilemmas, the formal theory of quantum mechanics has been immensely successful in predicting, describing, and utilizing myriad phenomena in the world of chemistry, electronics, astrophysics, and numerous other fields. As a practical tool in physics and engineering, it is unassailable and invincible. Among workaday physicists who benefit from the practical fruits of quantum mechanics, there is an eerie conspiracy of silence amounting to philosophical Philistinism about its meaning. Yet submerged within the seemingly secure foundations of quantum mechanics lie many enduring questions. The well-known remark of Einstein that "God does not play dice with the world," is the motto of the vanishing breed of physicist and philosopher who philosophically refuses to accept quantum mechanics at face value. But the contest of interpretations may be nearing an end. In recent years a series of experiments has suggested that, indeed, Nature not only hides behind a veil of uncertainty, but that until it is disrobed there is nothing behind the garment. Nature is something of a ghost—an exhibitionist one. It's in show business and needs an audience.

Unreality Unmasked

Not all physicists have accepted the slippery nature of reality—or should we say unreality?—suggested by the Copenhagen interpretation. As early as 1935 Einstein and his colleagues Boris Podolsky and Nathan Rosen challenged the interpretation with a "thought experiment," called, after its creators, the EPR paradox. They tried to demonstrate that quantum mechanics (QM) gave an "incomplete" description of nature, that beneath the seemingly *acausal* statistical behavior must lie a deeper mechanism—so-called hidden variables or hidden parameters. The EPR thought experiment suggested that QM led to a violation of Einsteinian "locality," the concept that event A cannot influence event B if they are so far separated that a signal traveling at the speed of light can't tell B about A before B has occurred. Einstein and his young colleagues said that because quantum mechanics seemed to allow instantaneous "spooky actions at a distance" in violation of locality, it must not be a thorough explanation of reality.

EPR was in the beginning literally a "thought experiment"—there existed no experimental methodology or apparatus to test it. Then in 1964 physicist John S. Bell developed a theorem, since called Bell's Inequality, that shows how the results of a generic class of physically possible EPR experiments can be interpreted mathematically to decide whether quantum mechanics is complete or if hidden variables are acting beneath the stage of quantum unreality. In the 1970s a series of EPR-like experiments employing photons of light, and in some cases high-energy gamma rays, firmly supported the conclusion that the quantum mechanical interpretation is valid. It seemed like the death knell for theories that deterministic hidden variables were pulling the strings of quantum puppets and giving the world an indecisive look.

In each of the experiments with low-energy photons, atoms

were energized to emit pairs of photons that flew apart in opposite directions. The precise geometric orientation of each photon's "vibration," called its polarization state, could be measured by detectors placed meters apart at opposite directions from the atomic source. Since the photons in each pair were known to have originated in an atomic process that made their polarization states *always* oppositely correlated, measuring the polarization of one of the pair effectively measures the other photon. (Simplifying a bit and ignoring the subtleties of the actual experiments, when photon A is "plus," photon B is "minus.")

The classical reality that Einstein and others wished to impose on such an experiment is that if A is measured "plus" and B of course is then logically "minus," this means that A was "plus" and B "minus" as they left the atom and up to the time of detection. Quantum mechanics, on the other hand, says that A doesn't become "plus" or B "minus" until the detector *measures* A as "plus." Because of the random way the measurements are made, this makes the quantum mechanical interpretation equivalent to a ghostly action at a distance; measuring A instantaneously seems to affect B across a physical gap unbridgeable by any signal (traveling no faster than light) in so short a time.

Bell's Inequality provides a remarkable statistical test of the subtle but profound distinction between the Einsteinian hidden variable viewpoint and the QM interpretation of this type of correlation experiment. Since experiments can't be performed with a single photon pair, millions of photons must pass through the polarization detectors to build up adequate correlation statistics. Unfortunately for those with Einsteinian psyches, the results of the EPR-like experiments done in the 1970s supported the quantum mechanical viewpoint, but there were still a few subtle objections to the way the tests had been conducted.

In 1982 physicist Alain Aspect and his colleagues in France

overcame criticisms of the earlier experiments with new refinements. These guaranteed that the two quantum measuring devices which were separated by many meters could not influence one another by signals traveling even as fast as light. The experiments make the case for quantum unreality seem ironclad. It is now nearly impossible to believe that hidden variables are at work in the quantum world. The microcosm is profoundly indeterminate until it is probed by a conscious observer. No one can satisfactorily explain why conscious witnesses—macroscopic beings themselves comprised of countless quantum particles—should have anything to do with the behavior of individual wave particles. As classical objective reality slips beneath a sea of quantum confusion, we grasp at straws to rescue it. Our efforts are in vain. Converted reluctantly, we must learn to live with an unfamiliar vision of the world's deepest mechanism, but how to come to terms with this blasphemous quantum weirdness?

A New Reality

First, have we overlooked any convenient escape routes? A few tempting paths do present themselves, but when following these avenues in search of reality, we find we are assaulted by a thicket of even more painful brambles. One path—the most alluring—is to believe that the world is pervaded by "nonlocal" influences. We imagine that every particle in the universe instantly influences every other, regardless of what we have been told about Special Relativity's prohibition on faster-than-light signals. We announce victoriously an opening at the end of the path: this must be the explanation of the Aspect experiments and their precursors.

Side benefits of our excursion into "Non-Locality" suddenly present themselves as true possibilities: telepathy, precognition—all manner of "psychic" phenomena! The meaning of causality for macroscopic events is destroyed. Alas, Non-Locality was only an evaporating dream. We find that the only non-local signaling that occurs in any of the experiments is transmission of random noise. It is pseudo faster-than-light signaling, not really sending any information at all. These are non-local influences far different than we had imagined. So we did not even restore objective reality by paying the high price of accepting bogus spooky actions at a distance. Despondently, we return to follow another path.

This time our course leads to a bizarre Alice-in-Wonderland world of illogic. Remembering the underlying assumptions of the EPR experiments, we throw caution to the wind and announce that quantum mechanics must not obey the ordinary (so-called Boolean) logic we have been accustomed to apply when we interpret scientific experiments. The very grammar of our feeble minds is turned on end. Relief from unreality is temporary indeed, as we realize that the new brand of logic, patched together to restore objective reality in the experiments, makes the entire scientific edifice tremble and come crashing down. Running back along the path of Contrary Logic, we emerge into the clearing and look for another escape.

And there it is! A path that leads inexorably to a guaranteed, consistent explanation of quantum mechanics! The path is labeled the Multi-Worlds Hypothesis, but it is a very strange way. At every quantum mechanical measurement we make along this route, the entire universe diverges into branches, each representing possible discrete quantum states. So all possible progressions of cosmic history, down to the most minute detail, exist simultaneously in parallel universes that are connected on an infinitely branching tree! Trillions upon trillions of universes appear each second, none of which we have any possibility of ever knowing a thing about. An extraordinary realization: many prominent physicists have been seen wandering down this path, blissfully admiring the foliage. The intellectual baggage that they carry along on these jaunts, seemingly oblivious to its weight, is awesome. Because buried in the apparent simplicity of the Multi-Worlds view, admittedly impossible to disprove, is an ugly complexity of parallel existences. It affronts the traditional model of simple scientific explanation, Ockham's Razor: given a choice, reject hypotheses that are most complex.

There are no more paths to explore, but weary from frightening journeys, we rest. Awakening from our dream about long-dead Objective Reality, we realize that we can simply choose to accept a New Reality, much as we did in accommodating the new reality of relativistic time and gravitation. Nature may be her ghostly self in the microcosm, but we still can expect proper conduct in the crude realm of large objects. Indeed, we are assured by quantum mechanicians that it is possible to view the world this way: an erratic, capricious microcosm, overlaid by a coarse world in which the Moon *is* there when we don't look at it and in which the tree *does fall* in the forest even when we (or our instruments) are not there to watch or record. The reason the macroscopic realm is more tangible and real may have something to do with the venerable "arrow of time." Memory of events and measurements can be stored in irreversible macroscopic processes—our own brains are good examples. But the quantum world has no memory. It doesn't truly exist until we larger beings touch it.

When all has been said, nothing will provide complete assurance that our New Reality is satisfactory. It still seems to be built on unreal quicksand that perpetually tries to consume us. Physicist Richard Feynman, a major contributor to the development of quantum mechanics, with his inimitable impish twinkle, boldly asserts, "I think I can safely say that nobody understands quantum mechanics." Niels Bohr said, "Those who are not shocked when they first come across quantum mechanics cannot possibly have understood it." And Einstein's eternal doubt, expressed in a letter to a friend in 1942, continues to haunt us: "It seems hard to sneak a look at God's cards. But that he plays dice and uses 'telepathic' methods (as present quantum mechanics requires of him) is something that I cannot believe for a single moment."

How does it happen that our biologically evolved brains can encompass, albeit with great effort, vastly different shades of reality, an ability that seems to surpass any reasonable requirement of animal survival? The multiplicity of quantum mechanical reality is but one illustration of the brain's uncanny and unreasonable effectiveness in comprehending as much as it does of the cosmic order. It brings to mind one of the greatest mysteries in cosmic evolution, perhaps the final enigma: Mind itself.

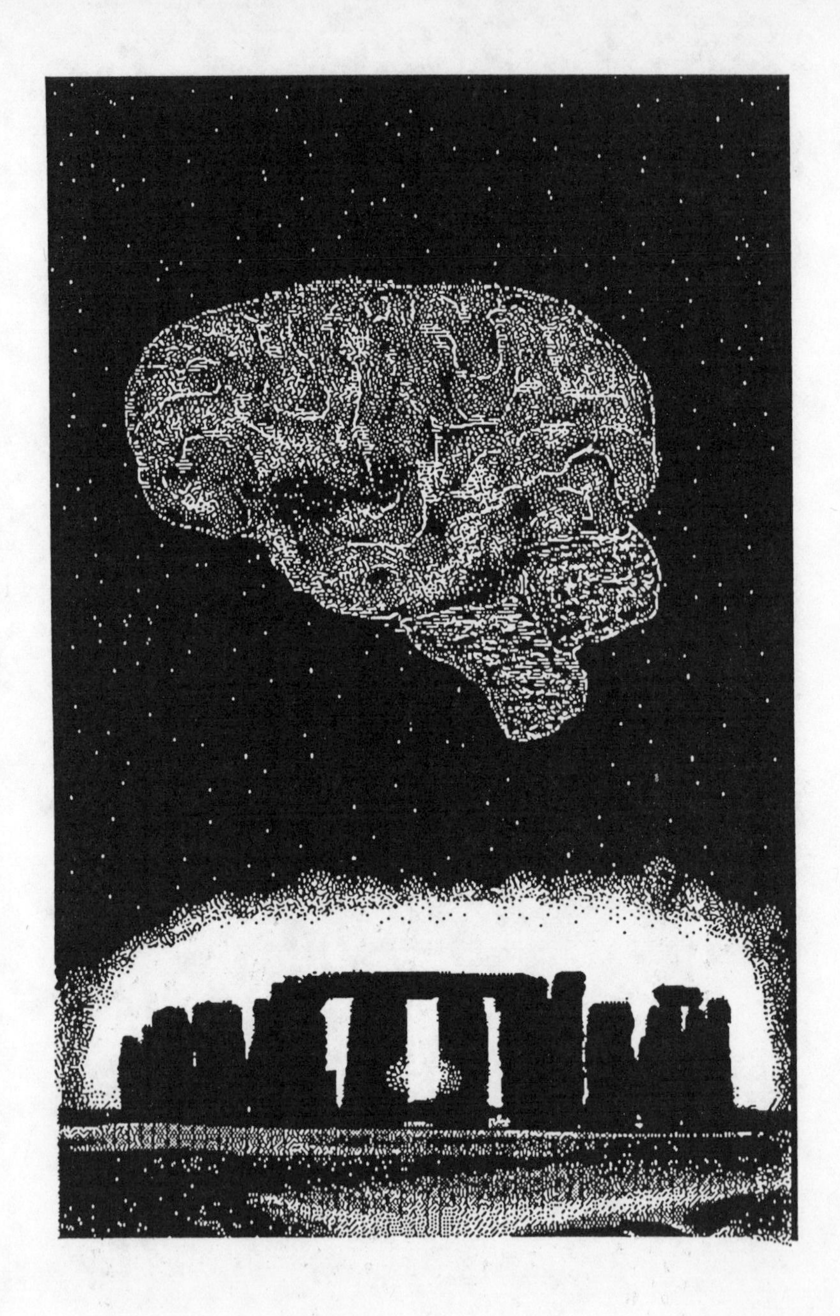

16
THE TRANSCENDENCE OF MIND

With science we can transcend our intuitions, just as with electronics we can transcend our eyes and ears. To the question of how such transcendence can have arisen in the course of biological evolution I have no satisfactory answer.

—MAX DELBRÜCK, *Mind from Matter?*, 1986

The great mathematician fully, almost ruthlessly exploits the domain of permissible reasoning and skirts the impermissible. That his recklessness does not lead him into a morass of contradictions is a miracle in itself. Certainly it is hard to believe that our reasoning power was brought, by Darwin's process of natural selection, to the perfection which it seems to possess.

—EUGENE P. WIGNER, "The Unreasonable Effectiveness of Mathematics in the Natural Sciences," 1959

In a universe of surpassing wonders, three beguile us far more than the rest. Foremost is the child of cosmology—existence—the wonder of the Beginning. An instant beyond cosmic genesis, reckoned by the span of ages yet to pass, lies the next—the origin of life, complexity. All the more glorious, because organization must surely have appeared on many worlds beyond Earth. Finally, the wonder of Mind that pervades the animal kingdom on our planet and, we confidently assume by analogy, the biota of other cosmic domains. The dumb universe could have ceased its creative work after the first of these natural miracles, but it didn't. Instead, the cosmos worked overtime. Mind emerged from the fabric of matter, and aeons from now it may make all the difference in the world.

Of course, our own minds interest us the most, but we can't deny that mind of a kindred sort is clearly at work within many other species—from the lowliest insects through cetacean behemoths whose mysterious minds, for all we know, may in some respects surpass our own mental abilities. Biologist Donald R. Griffin of Rockefeller University is a prominent advocate of the view that mind, a general conscious awareness of the environment, permeates the animal kingdom. He writes, "It seems likely that a widely applicable, if not all-inclusive criterion of conscious awareness in animals is '*versatile adaptability of behavior to changing circumstances and challenges.*'" These versatile adaptations he copiously documents in his book *Animal Thinking*. The most endearing instance, though perhaps not the most amazing, is the habit of the sea otter to select a rock of just the right heft to smash open shellfish on its abdomen while blissfully floating on its back. It may even swim for long periods with its favorite stone safely under arm. From examples like these, we have little doubt that mind is a property of so-called lower forms of life, but that it manifests more as a difference in degree rather than in kind.

Some scientists consider the emergence of advanced intelligence in the human species to be a quirk of evolution, an event that is likely not to be repeated often within organisms on other worlds. Since no one knows precisely what caused the species *Homo sapiens sapiens* to veer in the direction that it has, this probability game is so far all educated speculation. What seems for some to be a doubly contingent improbability—the genesis of life and the evolution of intelligence—appears to others to be a forced process, *especially* when the adaptive creativity demonstrated by less complex organisms is considered. The alleged improbability of life both originating and then attaining a modicum of advanced intelligence paradoxically makes the "assumption of mediocrity" all the more compelling—the hypothesis of a forced process toward life and intelligence at work throughout the cosmos. I am inclined to believe that the phenomenon of complexified mind is universal. The seeds of mind, after all, were planted more than 500 million years ago, perhaps much earlier. The first neuronal networks formed in the sea and later climbed with agonizing slowness onto the land. In fact, Lynn Margulis and Dorion Sagan suggest in *Microcosmos* that human and animal brains can be seen fundamentally as the product of "highly evolved bacterial association."

Note the implicit erasure of the ancient so-called mind-body or mind-brain problem from this cosmology, with due apologies to René Descartes and his categories of worldly substances *res cogitans* (mind) and *res extensa* (matter). Not because the historical development of the mind-body question isn't illuminating in its own right, but I personally feel no need to attribute, for example, a vitalist or nonphysical "spirit" that hovers within or outside a human brain. The music of ethereal matter seems spirit aplenty for the tapestry of mind, deep as matter may well be according to recent theories with vibrating strings of pure geometry embedded in a 10-dimensional space-time.

Consciousness is a phenomenon remarkable enough to have given rise to the plethora of legends about its intangibility. But it is easy to see how consciousness, willful as is its nature, can trick itself into anticipating its own immortality via a spirit world. Consciousness often aches to be relieved of its imprisonment in matter. In fact, a characteristic of consciousness—certainly *human* consciousness—is the fear of death, the acute, stabbing horror, upon waking from a peaceful afternoon nap, that someday all one's thoughts will end. Perhaps that fear is a fortunate evolutionary adaptation, until we learn to outgrow it; otherwise deadly carelessness or mass suicide would have terminated sentient life long ago. Though we don't know exactly how it works or how it evolved, we can describe consciousness as an emergent phenomenon within the complex organ called the brain. With no intention to debase the wonder and mystery of consciousness, it is *emergent* in a way analogous to meaningful television pictures arising from chaotic patterns of electrons that bombard a phosphorescent screen.

What we call conscious awareness is certainly far removed from the elemental electrochemical signaling transmitted through layer upon layer of cells to the neocortex. The emergent image of our surroundings—for example, colors and persistence of invariant geometric relations among objects—is a demonstrably higher order of mental processing than a mere two-dimensional retinal image. Neuroscience inexorably penetrates the linkages between primitive sensory inputs and higher level comprehension, and the brain becomes less mysterious though no less remarkable. Witness the extraordinary and rare phenomenon of synesthesia—literally, a joining of the senses. Synesthetes may see colored blobs in front of their eyes upon hearing specific sounds or words; they may taste or smell and simultaneously feel nonexistent geometric forms or experience myriad other distortions of the sensorium.

It seems that the evolution of neurological abilities that let animals cope with their environment has conspired to give rise in humanity to a surfeit of mental abilities, to which we have often alluded. There is no particular reason that we should be able to understand what we dangerously call the physical "laws of nature," or the mathematics that mysteriously develops in a timely fashion to illuminate them. There are other spectacular examples: visual art, poetry, and music. But are we the "idiot savants" of the universe, who, while manifesting certain profound talents, nonetheless remain oblivious to even deeper realities? Isn't the ultimate question we face: what are the limits of the knowable?

Knowability and the Unknown

Simply beholding the immense canopy of the heavens on a starstruck night is sufficient to convince us of the infinity of knowledge that we shall never possess. Were we to visit a different planet every second and miraculously absorb its history, beauty, the music, art, and philosophy of its possible inhabitants, we would spend more than ten thousand times the present age of the universe before we had completed our tour of even the present visible universe. The matter of an entire world converted to libraries would not suffice to store such immense knowledge.

Of course, we don't need to leapfrog from planet to planet; at the moment we would be deeply grateful to eavesdrop on a single sentient world for a few minutes. The example simply brackets the pathetic short reach of our vaunted grasp of the universe. Yet concerning cosmic knowledge, it is quality, not quantity, that we desire—universal generalizations rather than specifics, hence the appeal of striving to understand physical laws and mathematical "truths." But there is a dawning awareness that aspects of even these cosmic truths may permanently elude us. Perhaps this very confession of permanent ignorance about some things is the highest truth that we will ever know. We transcend the ordinary limitations of mind simply by being aware of them.

The late Max Delbrück, a Nobel laureate biologist and physicist, has written with enviable clarity about the borders of the known and the unknown, and how they came to be. He drew a distinction between *phylogenetic* and *ontogenetic* learning. The former is accomplished by the genetically evolved mental apparatus that facilitates survival as independent entities in the Darwinian milieu. Of phylogenetic learning he said, "What is *a priori* for the individual is *a posteriori* for the species." We literally know much about the world before we are even born; the potential knowledge is in the genes. Ontogenetic learning, by contrast, is culturally transmitted through a lifelong process of learning specific languages, mastering science, and absorbing innumerable other mnemonic structures.

It appears all but certain that our phylogenetic mental endowment establishes certain limits to what we can acquire ontogenetically. The paradigm of these obstacles is the proof that mathematician Kurt Gödel constructed in 1930. Thoroughly violating the prior intuition and expectations of eminent mathematicians like David Hilbert and John von Neumann, Gödel proved that within a framework of propositions (theorems) derived from a set of axioms (like the axioms of Euclidean geometry) there may be some propositions the truthfulness or falsity of which it is impossible to determine—i.e., *undecidable* propositions exist. Adding further to our sense of mathematical disillusionment, it is not even possible to determine whether or not a specific proposition is or is not decidable! Lying in wait within a set of axioms may be devilish contradictions ready to manifest themselves, but perhaps never to be discovered.

Not only does Gödel's proof insinuate doubt about the consistency and completeness of mathematics, it illustrates the limitations of mind with respect to the "real" world. So-called realist mathematicians, in the tradition of Plato, have long believed that mathematical relations were objective constructs—they seemed self-evidently to be cosmic truths. Gödel's proof tends to support the opposing "creationist" school that suggests that mathematical structures, as basic as the natural numbers used in counting, are mere constructs of our minds. Perhaps organisms evolved under different circumstances on other worlds (or on our own—e.g., cetaceans) would conceive mathematical labyrinths that are incomprehensible or contradictory to our human way of thinking.

Human beings have great difficulty, for example, comprehending the different degrees of infinity embodied in what are called *transfinite numbers.* And *prime* numbers (integers that are evenly divisible by no other number except themselves and one) are so elusive and mysterious that our minds wrap partially around them only with difficulty. Is it possible that organisms have evolved anywhere in the cosmos for whom infinities and prime numbers are as transparent and "obvious" as mundane numeration is to us? It seems unlikely, but that is only a terrestrial prejudice. Max Delbrück posed a question that touches on these mathematical limits of mind: "When we try to extend the use of numbers to realms that transcend our experience, or try to follow their implications too far and encounter paradoxes and contradictions, it is not so irrelevant to ask whether it is the mind or the world that lacks consistency."

Physicists may casually dismiss the internecine struggles of realists and creationists in the rarefied atmosphere of pure mathematics, but it seems urgent to ask whether physical theories themselves have a significant connection with reality. Or are they simply misguided though practical approximations to a world that human beings will never be able to comprehend because of the limitations of mind? The Newtonian framework of classical mechanics that seems ideally matched to human imagination overcame the comfortable Aristotelian perspective. Then relativity superseded the classical theories, but, nonintuitive as it may be, relativity is nonetheless comprehensible in a formal sense. Quantum mechanical reality is less formally understandable than relativity—witness the continuing philosophical battles over it. It seems inevitable that we will increasingly confront the outermost limits of mind as succeeding generations of theories approach and then go beyond gravity's rainbow—the unification of all the forces of nature.

Max Delbrück said that the conflict between external and internal reality is an illusion. In his cosmology, "The Cartesian cut between observer and observed, between inner and external reality, between mind and body, is based on an illusion that the physical world has no subjective component." If we believe Delbrück, we are then forced to conclude that different kinds of consciousness "create" different kinds of universes. Indeed, this seems to be the case within human culture. Perspectives on established theories (e.g., quantum mechanics) are not universal even in the community of science. And Delbrück believed that Jean Piaget's findings on the mental development of children suggest that infants have an image of the world closer to quantum mechanical reality than do adults with their hardened "scientific" belief in objective reality.

The transcendent quality of mathematical edifices and physical theories is their power to encapsulate in compact form things that appear to be large truths. Economy of expression for some reason strikes a pleasing chord in our minds, so we liken the search for truth to a quest for beauty. In the words of John Keats, "'Beauty is truth, truth beauty,' that is all ye know on earth, and all ye need to know." The mathematically *transcendental* number we designate with the Greek letter *pi* is beautiful, in part, because it permeates mathematical and physical theories, and because of the legendary mysteries of the infinite trail of digits that expresses it. But how the mind came to be tuned to this kind of beauty, and by natural extension to scientific truths, is a mystery that we are only beginning to fathom.

Philosopher Charles Hartshorne suggests that natural selection will favor imaginative animals which can deal with spontaneity in their environment. He speculates that "Animals must be able to *enjoy* imagining [my italics]; and then natural selection may favor those which have unusual amounts of this capacity and hence behave in unusually flexible ways. There must be an inner reward of imagination distinct from its eventual utility." Hartshorne concludes that, "The beauty of hypotheses, we may say, is the psychological ground of their possibility." Is this a partial answer to Delbrück's query about how science manages to transcend our intuitions? Did the sheer enjoyment of imagining, evolved perhaps from the thrill of the hunt or the manufacture of tools, engender the explosive evolution of ideas—the much wider hunt for beauty in the cosmos? As we often ask in science, is this idea really "beautiful enough to be true"?

An important aspect of mind that we usually overlook is its evolution and growth right before our eyes. Simply compare the larger body of knowledge that young children today are expected to acquire with what their counterparts two hundred years ago needed to master (not in every case a positive change!). Now this is ontogenetic learning evolved and handed down through written and now electronic records. But if we travel back 10,000, 20,000, then 50,000 years, we certainly expect a considerable change in phylogenetic learning abilities. Astronomer Gerald Hawkins, who decoded the mystery of Stonehenge, the 4000-year-old circular monument of stones on Salisbury Plain in southwest England, suggests something of that sort. He believes that the mental feats of memory and calculation necessary for Stone Age people to have built this lunar-solar calendrical device without benefit of written records, represents a mental prowess moderns no longer seem to rely on. Only a few geniuses or high priests may have possessed these mental abilities. So through the ages we have examples of the mind's ascent in some ways and atrophy in others. This is probably not a biological change in brain structure but a shift in emphasis in the mind's focus.

By imagining our conceptual framework of the universe to be the final word or nearly so, we risk deluding ourselves, much like the builders of Stonehenge may have. Mind, indeed, has wondrous powers to transcend the natural order and derive deeper theories that are in no way immediately apparent. But it also has an extraordinary capacity for self-delusion in many realms that include not only science and mathematics, but especially those sensitive fields we are warned to avoid in conversations with strangers—politics and religion. If there is an *absolute* "truth and beauty," how near does the quickening universe—a cosmology created by mind—approximate that patently unknowable Universe? Mind may at least aspire to this ultimate transcendence, though it will perhaps not be achieved even in that far-off time when noumenous ethereal beings may haunt the remains of an inconceivably barren cosmos. What role is there for the concept of God—the frequent principal of metaphysical cosmologies? And what significance, if any, can we attribute to the death of our own tiny physical essence, the end game of mind? A tentative answer comes from a very gentle soul who transcended an antique universe with the uncanny power of mind.

17

EINSTEIN'S GOD

I want to know how God created this world. I am not interested in this or that phenomenon, in the spectrum of this or that element; I want to know his thoughts; the rest are details.

—ALBERT EINSTEIN, 1955

Whoever undertakes to set himself up as judge in the field of Truth and Knowledge is shipwrecked by the laughter of the gods.

—ALBERT EINSTEIN, 1953

There are those who feel more deeply over religious matters than they do about secular things. It would be almost unbelievable, if history did not record the tragic fact that men have gone to war to cut each other's throats because they could not agree as to what was to become of them after their throats were cut.

—WALTER STACY, CHIEF JUSTICE OF THE NORTH CAROLINA SUPREME COURT, 1930

The practice of science is an often narrowly focused and highly reductionist enterprise. Painstaking analysis and reexamination of assembled facts follow time-consuming and difficult gathering of data. Yet, when the fine underground roots of scientific discovery come together, they join in a massive tree trunk of unified knowledge that magnificently expands outward. Without roots there would be no tree. The scientist finds joy in marveling at the tree even while being absorbed in its roots, and this is the bright confluence of science and what Albert Einstein called "cosmic religion." The hard work must be done, the sharply focused "spectrum of this or that element" must be gathered before there can be the privilege of stepping back and admiring.

If the quest for scientific wonder is deemed a godly pursuit, then science must surely be one of the most revealed religions. But its insights and prophecies arrive not from beautiful poetry in ancient texts, but in torrents of little revelations raining down—the daily reaffirmations of the cosmic order that scientists gather as they test the world. That is poetry in its own right. Science has many "prophets," but none whose vision of nature seems more congenial to a quickening universe than Albert Einstein. He stood on the shoulders of Isaac Newton and in a sense may have seen the body of God—beyond dead, absolute time and space to a vibrant, living universe of space-time tortured and twisted by matter.

Because Einstein, the humble and intensely private seeker of truth, was so reluctant to broadcast his views, few know him as the God-intoxicated man that he was. His writings and spoken words, sprinkled along the path of his seventy-six years, reveal an intensely religious person. While he lived, paradoxically some reviled him as an atheist, but they apparently didn't understand. In some ways Einstein has suffered the same fate of his seventeenth-century intellectual forebear, philosopher Benedict Spinoza, a similarly God-obsessed and misunderstood maverick.

Einstein's God was not the personal God of western religions, nor did his theology match religions of the Orient. He spoke and wrote of having a "cosmic religion," beliefs that he claimed were difficult to describe to anyone who is entirely without them. Central to his religiosity was, in his words, "a rapturous amazement at the harmony of natural law, which reveals an intelligence of such superiority that, compared with it, all the systematic thinking and acting of human beings is an utterly insignificant reflection." He did not believe in a personal God, writing in his 1931 essay "The World as I See It," "I cannot conceive of a God who rewards and punishes its creatures, or has a will of the kind we experience in ourselves.

Neither can I nor would I want to conceive of an individual that survives his physical death."

Mystery, but not untutored mysticism, was key to Einstein's religious sentiment. In words impossible to paraphrase without doing them an injustice, he wrote, "The most beautiful experience we can have is the mysterious. It is the fundamental emotion which stands at the cradle of true art and true science. Whoever does not know it and can no longer wonder, no longer marvel, is as good as dead, and his eyes are dimmed. It was the experience of mystery—even if mixed with fear—that engendered religion. A knowledge of something we cannot penetrate, our perceptions of the profoundest reason and the most radiant beauty, which only in their most primitive forms are accessible to our minds—it is this knowledge and this emotion that constitute true religiosity; in this sense and in this sense alone, I am a deeply religious man."

Einstein spent his early years as a nonpracticing Jew in Munich, Germany, aware of his heritage, yet in a family so assimilated and devoted to practicality that for sheer convenience young Albert went for five years to a Catholic elementary school. In his own autobiographical notes he described having attained a "deep religiosity" by age twelve. Having studied violin since age six, he also was struck by chords of musical influence. Ronald Clark, one of Einstein's biographers, speculating on the origin of his religious feelings, wrote, "Always sensitive to beauty, abnormally sensitive to music, Einstein had no doubt been impressed by the splendid trappings in which Bavarian Catholicism of those days was decked out."

Einstein's scientific career started at age five when his father showed him a pocket compass. Einstein later remembered wondering what invisible force could make the needle always point in the same direction. He went on to unify electromagnetism and mechanics with a consistent mathematical framework. By 1905 Einstein had published his theory of Special

Relativity that was soon to bring him nearly universal scientific acclaim. His 1916 General Theory of Relativity described gravity's origin in the curvature of space and time by matter, and through its breathtaking, imaginative insight catapulted Einstein to the Mount Parnassus of science. He spent the rest of his life in a failed attempt to unify the other known forces of nature with gravity, a task that seems much closer to realization thirty years past his death.

The revelations of his theories no doubt strengthened Einstein's belief in the paramount importance of comprehending the natural order. He made statements about this, which, taken out of context, might be mistaken for more conventional religious beliefs, e.g., "I want to know how God created this world. . . ." Einstein's God was the Universe itself, not a transcendent "grand puppeteer." And he had no doubt that there was a Universe, a deep, superpersonal reality, beyond the solipsism often so deceptively attractive to the human mind. He wrote in 1941, "A person who is religiously enlightened appears to me to be one who has, to the best of his ability, liberated himself from the fetters of his selfish desires and is preoccupied with thoughts, feelings and aspirations to which he clings because of their superpersonal value." And, "A religious person is devout in the sense that he has no doubt of the significance and loftiness of those superpersonal objects and goals which neither require nor are capable of rational foundation."

Einstein was misunderstood by religionists of varied persuasions, because, lacking scientific understanding, they could not see that the old-physics world of "simple" matter dispersed in vacuum had been replaced by a modern physics in which things are "not what they seem." Atoms are not hard little balls, and the "void" is not dull nothingness. Physics had grown far beyond rank materialism to embrace a pulsating, labyrinthian quantum world alive with energy and as ethereal as any heaven. In the words of physicist Edward Harrison, ". . . in the impalpable and seemingly inconsequential entities of the quantum world, one finds the true music and magic of nature." And Max Delbrück wrote, "Mind looks less psychic and matter less materialistic. . . ."

Einstein himself disparaged the "naïve realism" with which some still view the world: "This more aristocratic illusion concerning the unlimited penetrative power of thought has as its counterpart the more plebian illusion of naïve realism, according to which things 'are' as they are perceived by us through our senses. This illusion dominates the daily life of men and animals; it is also the point of departure in all of the sciences, especially of the natural sciences."

Einstein did not believe that science would ever know all that could be known about the world. As he confided in a friend, "Possibly we shall know a little more than we do now. But the real nature of things, that we shall never know, never." This point is the major theme of cosmologist Edward Harrison's book *Masks of the Universe*. Harrison eloquently traces humanity's quest to understand the world, and says that in every age humanity's world model or "universe" was thought to be the real Universe. There was the "magic universe" of prehistory in which the animism of all objects formed a continuum with living beings. This gave way to a succession of mythic universes with multiple powerful gods as prime movers. Thence to a medieval universe and a succession of physical universes. Harrison suggests that we shall never know the true "Universe," no matter how we embellish our transitory universe. He stands with Einstein in believing that we will never be able to attain complete knowledge of the Universe.

Though we might struggle 10,000 years to fathom the Universe and still not succeed, the quest is worthwhile. Einstein had faith that "God is subtle, but he is not malicious." By this he meant that even though the Universe did not reveal its inner workings easily, it held no absolute secrets. Indeed, he was impressed with its comprehensibility. After all, one could imagine a chaotic world without rhyme or reason, a world impossible to understand by any simple set of laws. But the world is far from that way—it is strikingly regulated. Einstein believed that faith in this regularity came from "religion": "Science can only be created by those who are thoroughly imbued with the aspiration toward truth and understanding. The source of this feeling, however, springs from the sphere of religion. To this there also belongs the faith in the possibility that the regulations valid for the world of existence are rational, that is, comprehensible to reason. I cannot conceive of a genuine scientist without that profound faith. The situation may be expressed by an image: science without religion is lame, religion without science is blind."

The comprehensibility of the world was a wonder to Einstein. It was a hallmark of nature that he pointedly used against stiff-necked atheism. In a letter to a longtime friend in 1952, Eisntein wrote, "And here lies the weak point for the positivists and the professional atheists, who are feeling happy through the consciousness of having successfully made the world not only god-free, but even 'wonder free.' The nice thing is that we must be content with the acknowledgement of 'wonder' without there being a legitimate way beyond it." There were no final answers to the ultimate questions of science, a humble realization he thought could moderate the turbulence of human conflicts brewing about him. In a 1932 letter to Queen Elizabeth of Belgium, whom he knew as a friend, he wrote, "One has been endowed with just enough intelligence to be able to see clearly how utterly inadequate that intelligence is when confronted with what exists. If such humility could be conveyed to everybody, the world of human activities would be more appealing."

Einstein had not yet moved to the United States to escape the coming European nightmare, but the magazine section of

The New York Times of November 9, 1930, featured an article by him entitled "Religion and Science." In it he discussed his "cosmic religion" and its relation to science and other varieties of religious experience. According to Ronald Clark, Catholic theologian Dr. Fulton Sheen called it the "sheerest kind of stupidity and nonsense. There is only one fault with his cosmical religion: he put an extra letter in the word—the letter 's.'" A New York rabbi, Nathan Krass, averred, "The religion of Albert Eisntein will not be approved by certain sectarians, but it must and will be approved by the Jews." A few years before this episode, Cardinal O'Connell of Boston had said to his audiences that Einstein's General Theory of Relativity (the good cardinal presumably understood it well, after reading about it in *The New York Times*) was "cloaked in the ghastly apparition of atheism." That prompted another New York rabbi to seek assurances from Einstein, leading to the famous reply, "I believe in Spinoza's God who reveals himself in the orderly harmony of what exists, not in a God who concerns himself with fates and actions of human beings."

Nearly 300 years before Einstein's flirtation with religious controversy, Baruch Spinoza was born to a family of Portuguese Jews in Holland. The family had emigated rather than face the forced conversion of the Inquisition. Baruch was steeped in the learning and history of his people's Diaspora that began 1500 years earlier. He had a lifelong Jewish heart. But his own critical appraisal of the Bible and the influence of freethinking that came out of the Renaissance sealed his fate. The elders of his synagogue charged that he was saying that "God had a body—the world of matter; that angels might be hallucinations; that the soul might be life itself; and that the Old Testament did not affirm an afterlife." Since he would not recant, he was excommunicated. According to Will Durant, the elders felt "that gratitude to their hosts in Holland demanded the excommunication of a man whose doubts struck at Christian doctrine quite as vitally as at Judaism." In exile, he changed his name from Baruch, a Hebrew word meaning blessed, to Benedict. Spinoza penned lines that apparently were so congenial to Einstein, "The philosopher knows that God and nature are one being, acting by necessity and according to the invariable law; it is this majestic Law which he will reverence and obey."

In *Masks of the Universe* cosmologist Edward Harrison surveys many philosophies of existence and adopts the Spinoza-Einstein view. He writes that his ideas come from "agnostic soil," though he has a Protestant background. He encapsulated the modern religious dilemma, "Rejection of the possibility of a God-Universe or UniGod perhaps explains why we find ourselves in desperate need of proof of God's existence. Long ago human beings abstracted from the natural world all that they ascribed to the gods, leaving the world dead; now the gods have fled into a surrealistic world of improbable existence, taking with them the half of the natural world that we call divine. We ourselves have transformed God into a fiction that cannot be proved true."

Though Albert Einstein did not believe in the creating and fostering God of the Bible, he had profound respect for what he called "religious geniuses" who revealed moral conduct to humanity. Einstein realized the limitations of science when he wrote, "Science can only ascertain what *is*, not what *should be*." He did not believe that reason alone could generate moral imperatives. He said fundamental ends "exist in a healthy society as powerful traditions, which act upon the conduct and aspirations and judgements of the individuals; they are there, that is, as something living, without it being necessary to find justification for their existence." And from where did these moral codes derive? According to Einstein, "They come into being not through demonstration but through revelation, through the medium of powerful personalities. One must not attempt to justify them, but rather to sense their nature simply and clearly." Einstein's God was revealed in the laws of physics, but ethical principles he took from the sages of all religions.

The paradox of Einstein's achievements must be counted one of the supreme ironies of history. Here was the essential pacifist who despised militarism, yet whose theories helped to unchain the nuclear genie. Never a practicing Jew, he nonetheless had the greatest affinity for the Jewish people and their postwar redemption in Israel. He was offered the presidency of Israel in 1952 but respectfully declined it. In his last years he wrote, "As to my work, it no longer amounts to much. I don't get many results any more and have to be satisified with playing the Elder Statesman and the Jewish Saint, mainly the latter."

So Einstein's legacy must include not only his physical theories but his cosmic religion—little known and little shared until perhaps another age. He challenged the future: "I maintain that the cosmic religious feeling is the strongest and noblest motive for scientific research." And, "In my view, it is the most important function of art and science to awaken this feeling and keep it alive in those who are receptive to it."

Einstein's life ebbed and evaporated in a hospital bed in the early morning hours of April 18, 1955—victim of an (at that time) inoperable aortic aneurism. He mumbled his final words in German to an uncomprehending attendant. Perhaps the words paraphrased his earlier expressed sentiment, "Is there not a certain satisfaction in the fact that natural limits are set to the life of the individual, so that at its conclusion it may appear as a work of art?" His corporeal atoms were seared in the fires of cremation and were scattered, as he wished, where no monument could be built. The seat of Einstein's mind, his brain, was removed for study—a scientific monument of sorts. Yet this curious and lonely human being's spirit—if we dare call it that—lives on in the world, sans brain, sans body. Much cosmic business remains unfinished. . . .

Immortality? There are two kinds. The first lives in the imagination of people, and is thus an illusion. There is a relative immortality which may conserve the memory of an individual for some generations. But there is only one true immortality, on a cosmic scale, and that is the immortality of the cosmos itself. There is no other.

—Albert Einstein

18

The Immanence of Meaning

It is very hard to realize that this all is just a tiny part of an overwhelmingly hostile universe. It is even harder to realize that this present universe has evolved from an unspeakably unfamiliar early condition, and faces a future extinction of endless cold or intolerable heat. The more the universe seems comprehensible, the more it also seems pointless.

—Steven Weinberg, *The First Three Minutes*, 1977

I would rather think of life as a good book. The further you get into it, the more it begins to come together and make sense.

—Rabbi Harold Kushner, *When Everything You've Ever Wanted Isn't Enough*, 1986

The profound fact of oneness not only on Earth but in all creation is becoming more and more evident as knowledge expands outward, and nowhere are to be found any absolute boundary lines or uncrossable barriers between any kingdoms of life. Not even between life and nonlife, nor between body, mind, and spirit. . . .

Are we then God's dream set to music in the place where the sea and the wind have begun to awake and think?

—Guy Murchie, *The Seven Mysteries of Life*, 1978

In some sense man is a microcosm of the universe; therefore what man is, is a clue to the universe. We are enfolded in the universe.

—Physicist David Bohm, in Renée Weber's *Dialogues with Scientists and Sages: The Search for Unity*, 1986

Up, up the delirious heights. We cannot rest. Seeking to transcend the boundaries of what can be known, too soon we lose contact with the toiling worlds of biology, physics, and mathematics. A bedazzled explorer must proceed with extreme caution, or be lost in a quagmire of meaningless sentiment. We have had a long journey from the edge of the sea to this mountain overlooking the infinite. It is impossible not to savor the totality of the immense vista—to enjoy the wholeness of it.

It isn't surprising to find that people who for one reason or another have little appreciation of science conclude that we live in a meaningless universe. Or alternately that they focus obsessively on a human-centered existence under the aegis of a transcendent father figure. The absurdities and injustices of imperfect human affairs, after all, leave much to be desired—and explained. But science is one of the few genuine paths to understanding, even the primitive origins and Darwinian evolution of human behavior that bear so many elements of tragedy, comedy, and hope.* Physicist Steven Weinberg writes cogently, "The effort to understand the universe is one of the very few things that lifts human life a little above the level of farce, and gives it some of the grace of tragedy."

Of what good is this unfinished pastorale, if with it we cannot satisfy a hunger for meaning? Yet some scientists, like Steven Weinberg, whom I also quote in the leading epigraph, claim not to have a craving for universal meaning—or having it, deny that it exists. A physicist may make brilliant leaps in penetrating the early stages of cosmic evolution, but does his "faith in meaninglessness" then represent conviction or perhaps genuine embarrassment? If conviction, for the depths of what reason? If embarrassment, may it not arise from outrages against rationality and justice that errant politics and religion have throughout history perpetrated in the name of righteousness? It may be unsettling to affiliate, even ambiguously, with former and present expropriators of meaning who know little of the beauty of science, and who care even less to approach it.

---

*See *Darwinism and Human Affairs* by Richard Alexander.

On the other hand, David Bohm, a respected pioneering physicist who has pondered the wider implications of physical theory, suggests that *meaning* is the essence of *existence*. This goes beyond physicist John Wheeler's quantum observers in his "participatory universe." Bohm says that the very act of questioning the nature of being—assigning meaning to the interplay of the limited aspects of the greater reality that we can perceive—is itself a *form of being*. It is the quickening universe coming alive, examining, and redefining itself. Philosopher Renée Weber describes Bohm's opinions in her book *Dialogues with Scientists and Sages: The Search for Unity* as "assigning a role to man that was once reserved for the gods." In Bohm's cosmology, human beings and other sentient entities inhabiting the universe are not gods, but they are part of God or Universe—call it what you will—in dialogue with itself.

Growing understanding of cosmic evolution increasingly merges ancient questions of "why" with those of "how." Because of our broadening perspective "how," in effect, *becomes* "why." The terms lose their distinction, faced with a Universe bearing perhaps an infinite regression of physical foundations—perhaps not a truly infinite staircase of causes, but infinite at least in terms of our inability ever to fathom them all. To categorize these views theologically, one might call them *pantheistic*—a tradition that extends back to the Greek Stoics. Pantheism regards God as the indwelling presence of the divine in all things; that is, the divine is *immanent* in all of nature. Pantheism is thus distinguished from classical *theism* that considers God to be *transcendent*—beyond nature. Another variant of belief, called pan*en*theism, considers the divine to have both immanent and transcendent qualities. While one can argue with much merit that these are mere words to disguise ignorance of the Universe, they nonetheless reflect attitudes toward Nature that are quite distinct.

A purely theistic viewpoint treats Nature as a footnote to a transcendent divinity, one clearly not to be approached intimately through science. Pantheism, on the other hand, becomes essentially identical with the content of science or all that science could ever hope to reveal. Panentheism seems to be more in tune than pantheism with modern scientific and mathematical discoveries—for example, quantum mechanical acausality and mathematical undecidability. Panentheism acknowledges the transcendent quality and unknowable essence of the Universe, while still preserving the role of the immediate experience of Nature offered by science. The theistic need for supernatural events—miracles—may simply represent a fundamental misunderstanding of and dissatisfaction with nature as it is, or should we say, as it is incorrectly imagined to be. Nature, viewed intimately with a deep sense of awe, not devalued as base, naïve materialism, is seen to express "miracles" continuously. And thus we come to the idea of *mysticism*—the experience of unified meaning immanent in all things.

Science and Mysticism

The feeling of oneness or unity with the world is the essence of the mystical experience. We can approach mysticism either from a scientifically uninformed state of pure emotion or from a perspective that draws mystical feelings from the scientifically revealed order. Mysticism is a much-maligned concept, particularly among scientists, most of whom believe it to be "other worldly," quite antithetical to their discipline. Conversely, self-proclaimed mystics are not notable for their appreciation of scientific subtleties. There are, however, dramatic exceptions on both sides. Renée Weber draws this comparison: "A parallel principle drives both science and mysticism—the assumption that unity lies at the heart of our world and that it can be discovered and experienced by man. *I believe that this one similarity is so powerful that it transcends the many differ-*

ences which divide science and mysticism." The late Jacob Bronowski, a mathematician and man of letters not prone to mysticism, nonetheless defined science as "the search for unity in hidden likenesses"—a mystical quest if ever there was.

The process of scientific discovery must ordinarily stand apart from the emotion of mysticism. One can't, of course, always be smelling the roses or breathing mountain air on the way to the scientific frontier—the grease of telescope gears and the pungent aroma of the laboratory more often accompany the scientific enterprise. But the *process* of science can be its own reward, apart from its practical benefits or the mystical enjoyment of an expanding image as another piece of Nature's jigsaw puzzle snaps sharply into place. Aside from a few remarkable turns where mystical wonder may have helped empower a scientific revolution, such as Kepler's synthesis of the laws of planetary motion, science can proceed without mysticism. Science has much to recommend it for its own sake, but I am not convinced that mysticism can stand honestly on its own. On the other hand, I recognize that this belief may derive from the limitations of an individual human experience. Often the dividing line between mystical feeling and scientific inspiration *is* very thin. For example, the notion of the unity of the biosphere—Gaia—is both scientific and mystical, as is the search for unity in physical theory. The idea of cosmic evolution, the search for extraterrestrial intelligence, a quickening universe—these too are iridescently scientific and mystic.

Mystical feelings give rise to a sense of wonder, much as do breathtaking scientific insights. Mysticism evoked by scientific thought needn't wait for the drama of turning points on the road to remarkable discoveries. The experience may occur in the course of both ordinary and extraordinary events of everyday life, the daily enjoyment of contemplating the stage of cosmic evolution in which we find ourselves. For example, when observing human beings walking to and fro, I marvel at their mobility and detachment from the Earth. I am flushed with awareness that part of the planet has after a long process liberated itself from its past molecular bondage and now moves free. On a clear night, gazing upward at the silent firmament, I am at once struck by its infinite depth and an awareness that but a short cosmic moment ago, the universe was collapsed within a pregnant void. A few cosmic ages beyond our time as the destiny of the cosmos unfolds, what the progeny of today's universe will have become is beyond imagining. The feeling is profoundly mystical, one of claustrophobia within infinitude, of transcendent destiny within insignificance.

There have also been those extraordinary moments coinciding with giant leaps in local evolution. We have been fortunate witnesses to many of them in our time, and more are sure to come. From the beaches of southern Florida on a humid July morning that is still green in memory, I watched a towering pillar of fire carry the first living emissaries of this world to another. A few days later the ancient dreams of quaking, Moon-watching life were fulfilled on the dusty Sea of Tranquillity. Time seemed to stand still in the mystical interlude as the Earth shrugged off a bit of life and then returned it to the fold. Leap ahead fourteen years to a tranquil New England village where primate technology has reached a climactic pinnacle—the inauguration of Project Sentinel, the most advanced effort to listen for the heartbeat of the quickening universe.

The Three Levels of Meaning

We have come alive in a cosmos that reflects meaning in every sinew, significance in every atom, yet we are free to ignore it entirely and spend our days immersed in a druglike stupor of superficiality. Of course, life is more than an exploration of meaning. It is for love and fellowship, for the pursuit of happiness, for play, and for the stress of survival, all of which are nurtured by human freedom and an absence of irrational dogmas that impose a sort of terrorism of the mind. But what a monumental waste if we should devote all our critical faculties, honed by aeons, completely to trivial pursuits—or worse, self-destroying ones.

Have we really come all this way for our descendants to know us in the fossil record as "homo superficialis?" Can we not as a civilization rise above the prevailing and deadly *Time* magazine mentality of instant analysis of transient events that are elevated to apocalyptic dimensions, then soon forgotten for next week's story? Where is cosmic perspective in our vaunted technological civilization? For that matter, where is cosmic awareness on the whole contending planet? Closeted in library and laboratory, no doubt, but certainly nowhere to be found in the highest councils of governments anywhere.

Ralph Waldo Emerson, the nineteenth-century American transcendental poet, was echoing Socrates and Plato when he spoke these stern words: "The unexamined life is not worth living." Extending and giving a scientific cast to that theme, we could say with equal force, "The unexamined universe is not worth living *in*." And since we are, as David Bohm suggests, "a microcosm of the universe," we must examine the meaning not only of cosmology and quickening, but also of humanity and the evolution of culture. The search for meaning thus proceeds at the levels of cosmos, civilization, and the individual.

The meaning I discern at the level of cosmos is humility in the face of the deep beauty and mystery that confronts us. As Einstein wrote, "Measured objectively, what a man can wrest from Truth by passionate striving is utterly infinitesimal. But the striving frees us from the bonds of self and makes us comrades of those who are the best and the greatest." One might say, kinship not only with the "best" human beings, but also with intelligent creatures on other worlds. The universe is but a dream, yet our minds seem to encompass part of it. We understand the reason for this very little, if at all, but we can aspire to an explanation. And what is the meaning of that plenum of nonexistence which "contained" the universe before it emerged? Is it perhaps disembodied mathematical laws, pure potentiality? What else could it be? It isn't satisfactory, but perhaps the closest we will get to the Universe is a mystical hypothesis made by Emerson in his 1836 essay, "Nature": "Nature is the incarnation of thought. The world is the mind precipitated."

On the time scale of the universe, terrestrial civilization teeters on the brink of thoughtless extinction—through neglect to *consider* not only the prospect of global nuclear immolation but also the distinct possibility that an unsuspected natural catastrophe could suddenly end our lineage. Contrary to popular opinion, species *Homo sapiens sapiens* does not lead the first charmed existence in the history of evolution. Of the many possible finales: perhaps an Earth-crossing asteroid of deadly-heft slamming into the planet and decapitating life, as one may have done in the age of dinosaurs. The projectile would be oblivious to proposed space defense shields, but not to infinitely less costly round-the-clock astronomical monitoring that could dispatch rockets in a timely fashion to divert the interloper, thus saving our skins and lineage. We recall that on the morning of June 30, 1908, a tiny asteroid or comet hurtled through the sky above the Stony Tunguska River in a remote area of Siberia, releasing the explosive energy of a 12-megaton hydrogen bomb. Eugene Shoemaker, a noted geologist and expert on asteroid impacts, estimates the probability of another Tunguska-like event in the next seventy-five years at 12 to 40 percent. Perhaps a city would be its victim next time. A much rarer larger body—say, 10 kilometers in diameter—would be required to devastate global civilization, but the calamity could happen at any moment. Or will humanity's final chapter be written by restless RNA or DNA, naturally transmogrifying into deadly viral pandemic? What if the AIDS (acquired immune deficiency syndrome) virus had been as readily transmissible as a flu virus? Are we on guard? Do we—as would be relatively easy by now—have colonies in space, on the Moon, or on the planets, to serve as ultimate safeguards to the extinction of human life? Of course not! We're too busy being at each other's throats and being immersed in stuporous ephemera, such as often mindless kinds of television fare and equally transitory journalistic obsessions.

What meaning can we assign to the ill-suited cosmic comic-tragedy, a distasteful opéra bouffe, that civilizations now play on the surface of our emerging world? Perhaps the message is that life has still much to learn. Human culture remains locked in a Darwinian survival stage where politics is the highest organizing principle. "Politics," Einstein aptly wrote, "is a pendulum whose swings between anarchy and tyranny are fueled by perennially rejuvenated illusions." Few make an effort to step back from illusion and cultural ephemera and see the forest instead of the trees. One conclusion is nearly certain: the all-too-easy extinction of human civilization would not be the end of life on Earth, nor even perhaps the end of putative advanced intelligence on this planet. If civilization should perpetrate the ultimate denial of cosmic trust, then slowly and painfully again, lesser life-forms would emerge to carry the fire. As I walk along the streets of Washington, I observe squirrels with glazed looks scampering earnestly about their nut-gathering business, not unlike the human pursuits surrounding me. The eerie thought arises that perhaps these ignorant yet guileless creatures—not the hardy microbes of Gaia—will eventually be heir to the globe. Perhaps the age of "*Squirrelo sapiens*" would dawn millions of years after the errant asteroid struck or the well-prepared thermonuclear funeral pyres blazed.

Could it be that civilization is in a stage much like the dying person, who, myth says, seconds before expiring witnesses in review all the events of life? Is our emerging picture of cosmic evolution this kind of recapitulation, mere terminal preparation? Evolutionary biologists have found in the fossil record that just before a species becomes extinct, it manifests frenzied activity—a burst of proliferation. But I am reminded that a woman in the final hours before onset of labor, also naturally enters a frenetic stage to prepare for the new life. I would rather believe that, like the plump and florid mother, civilization is anticipating a momentous rebirth instead of its demise. Perhaps with the paradigm of cosmic evolution in firmer control of the often fragmented scientific enterprise, science will be able to aid civilization through that rebirth. Science is conservative, yet is also a supremely subversive enterprise that casts all into doubt. Sometimes it can move mountains. One such mountain would be irrefutable proof that we are not alone in the galaxy. Or has the Doomsday clock crept too far? Perhaps it is hopelessly Utopian to imagine a world in which science is of major interest to political leaders for reasons other than practical utility or to destroy unfriendly tribes. But because we are human, we can still hope and strive for that day. We wistfully recall that leaders of Thomas Jefferson's and Benjamin Franklin's caliber had a keen appreciation of the science of their day.

And so we return to the edge of the sea whence we came, seeking meaning for ourselves in the waters. As we wade into the warm ocean on a star-dark night, crescent moonlight glistens dimly over the chaotic waves and follows us. We raise our heads to the sky and at once experience the beauty, the romance, the enfolding love and music of the universe in transition. In our mind's eye we see the antique message of the Psalmist that is emblazoned on an observatory in Vermont where seekers of stars sojourn in peace: "The heavens declare the glory of God." Surely there is beauty and meaning to be found even in the ancient poetry, celebrations, and rituals that elevate those who appreciate the song of the universe in another way. But whoever has patiently listened to the music of the cosmos, to the beating of atoms and the throbbing of stars, cannot turn back. They must take leave of myths past and seek a story for the future.

Our quest unfinished. Soon we shall hand down hard-won yet meager enlightenment to our clinging children, as those who went before bequeathed their knowledge to us. What has been our place here? Simply to seek truth and know beauty. This we take with us to the end: one atom in the quickening sea of a barren world can ignite the wildfire of life that consumes a globe, or one being on a germinating planet can alter the course of a civilization. One curious and romantic civilization may aspire to a galaxy. Civilizations of goodwill together may change a universe. There is meaning in the mystery of this story that can have no end.

FURTHER READING

Books

Alexander, Richard D. *Darwinism and Human Affairs*. Seattle: University of Washington Press, 1979.

Barrow, John D. and Joseph Silk. *The Left Hand of Creation: The Origin and Evolution of the Expanding Universe*. New York: Basic Books Inc., 1983.

Barrow, John D. and Frank J. Tipler. *The Anthropic Cosmological Principle*. New York: Oxford University Press, Inc., 1986.

Billingham, John, ed. *Life in the Universe*. Cambridge, Massachusetts: M.I.T. Press, 1981.

Bohm, David. *Wholeness and the Implicate Order*. London: Ark Paperbacks, 1980.

Bright, Michael. *Animal Language*. Ithaca: Cornell University Press, 1984.

Bronowski, Jacob. *The Ascent of Man*. Boston: Little Brown and Company, 1973.

Cairns-Smith, A. G. *Genetic Takeover and the Mineral Origins of Life*. Cambridge, Great Britain: Cambridge University Press, 1982.

———. *Seven Clues to the Origin of Life: A Scientific Detective Story*. Cambridge, Great Britain: Cambridge University Press, 1985.

Chaisson, Eric. *Cosmic Dawn: The Origins of Matter and Life*. Boston: Atlantic Monthly Press, 1981.

Churchland, Paul M. *Matter and Consciousness*. Cambridge, Massachusetts: M.I.T. Press, 1984.

Further Reading

Clark, Ronald W. *Einstein: The Life and Times.* New York: The World Publishing Company, 1971.

Crick, Francis. *Life Itself: Its Origins and Nature.* New York: Simon and Schuster, Inc., 1981.

Curtin, Deane W., ed. *The Aesthetic Dimension of Science.* New York: Philosophical Library, 1982.

Davies, Paul. *Other Worlds: Space, Superspace, and the Quantum Universe.* New York: Simon and Schuster, Inc., 1980.

———. *The Edge of Infinity.* New York: Simon and Schuster, Inc., 1981.

———. *The Accidental Universe.* Cambridge, Great Britain: Cambridge University Press, 1982.

———. *God and the New Physics.* New York: Simon and Schuster, Inc., 1983.

———. *Superforce.* New York: Simon and Schuster, Inc., 1984.

Dawkins, Richard. *The Selfish Gene.* New York: Oxford University Press, Inc., 1976.

Day, William. *Genesis on Planet Earth*, Second Ed. New Haven: Yale University Press, 1984.

Delbrück, Max. *Mind from Matter: An Essay on Evolutionary Epistemology.* Palo Alto: Blackwell Scientific Publications, Inc., 1986.

Drexler, K. Eric. *Engines of Creation: Challenges and Choices of the Last Technological Revolution.* New York: Anchor Press/Doubleday, Inc., 1986.

Dukas, Helen and Banesh Hoffman. *Albert Einstein: The Human Side.* Princeton: Princeton University Press, 1979.

Dyson, Freeman. *Disturbing the Universe.* New York: Harper and Row, 1979.

———. *Origins of Life.* Cambridge, Great Britain: Cambridge University Press, 1985.

Einstein, Albert. *Ideas and Opinions*. New York: Crown Publishers, Inc., 1954.

Eiseley, Loren. *The Immense Journey*. New York: Alfred A. Knopf, Inc., 1957.

———. *Darwin's Century*. New York: Doubleday and Company, Inc., 1958.

———. *The Unexpected Universe*. New York: Harcourt Brace Jovanovich, 1964.

———. *The Invisible Pyramid*. New York: Charles Scribner's Sons, 1970.

———. *The Night Country*. New York: Charles Scribner's Sons, 1971.

———. *The Star Thrower*. New York: Times Books, 1978.

Eldredge, Niles. *Time Frames: The Rethinking of Darwinian Evolution and the Theory of Punctuated Equilibria*. New York: Simon and Schuster, Inc., 1985.

Fagg, Lawrence W. *Two Faces of Time*. Wheaton, Illinois: Theosophical Publishing House, 1985.

Feinberg, Gerald and Robert Shapiro. *Life Beyond Earth: The Intelligent Earthling's Guide to Life in the Universe*. New York: William Morrow and Company, Inc., 1980.

Feinberg, Gerald. *Solid Clues: Quantum Physics, Molecular Biology, and the Future of Science*. New York: Simon and Schuster, Inc., 1985.

Feynman, Richard. *The Character of Physical Law*. Cambridge, Massachusetts: The M.I.T. Press, 1965.

Finney, Ben R. and Eric M. Jones. *Interstellar Migration and the Human Experience*. Berkeley: University of California Press, 1985.

Forward, Robert L. *Dragon's Egg*. New York: Ballantine Books, 1980.

———. *Starquake*. New York: Ballantine Books, 1985.

Gal-Or, Benjamin. *Cosmology, Physics, and Philosophy.* New York: Springer-Verlag, 1981.

Goldsmith, Donald and Tobias Owen. *The Search for Life in the Universe.* Menlo Park, California: Benjamin/Cummings Publishing Company, 1980.

Gould, Stephen Jay. *Ever Since Darwin.* New York: W. W. Norton Company, Inc., 1977.

———. *The Panda's Thumb.* New York: W. W. Norton Company, Inc., 1980.

———. *Hen's Teeth and Horse's Toes.* New York: W. W. Norton Company, Inc., 1983.

———. *The Flamingo's Smile.* New York: W. W. Norton Company, Inc., 1985.

Gribbin, John. *Genesis: The Origins of Man and the Universe.* New York: Delacorte Press, 1981.

———. *In Search of Schrödinger's Cat: Quantum Physics and Reality.* New York: Bantam Books, Inc., 1984.

———. *In Search of the Big Bang: Quantum Physics and Cosmology.* New York: Bantam Books, Inc., 1986.

Griffin, Donald R. *Animal Thinking.* Cambridge, Massachusetts: Harvard University Press, 1984.

Gunn, James E. *The Listeners.* New York: Charles Scribner's Sons, 1972.

Harrison, Edward. *Masks of the Universe.* New York: Macmillan Publishing Company, 1985.

Hartman, H., J. G. Lawless, and P. Morrison, eds. *Search for the Universal Ancestors,* NASA SP-477. U.S. Government Printing Office, 1985.

Hawkins, Gerald S. *Mindsteps to the Cosmos.* New York: Harper and Row Publishers, 1983.

Heisenberg, Werner. *Physics and Philosophy: The Revolution in Modern Science*. New York: Harper, 1958.

Herbert, Nick. *Quantum Reality: Beyond the New Physics*. New York: Anchor Press/Doubleday, 1985.

Hofstadter, Douglas R. *Gödel, Escher, Bach: An Eternal Golden Braid*. New York: Basic Books, Inc., 1979.

Hofstadter, Douglas R. and Daniel C. Dennett. *The Mind's Eye: Fantasies and Reflections on Self and Soul*. New York: Basic Books, Inc., 1981.

Hofstadter, Douglas R. *Metamagical Themes: Questing for the Essence of Mind and Pattern*. New York: Basic Books, Inc., 1985.

Hoyle, Fred. *The Black Cloud*. New York: Harper and Brothers Publishers, 1957.

Lovelock, James E. *Gaia: A New Look at Life on Earth*. Oxford, Great Britain: Oxford University Press, 1979.

Kolb, Edward W., Michael S. Turner, David Lindley, Keith Olive, and David Seckel. *Inner Space, Outer Space: The Interface Between Cosmology and Physics*. Chicago: The University of Chicago Press, 1986.

Kühn, Thomas S. *The Structure of Scientific Revolutions*, Second Edition. Enlarged. Chicago: The University of Chicago Press, 1962.

Margulis, Lynn and Dorion Sagan. *Microcosmos: Four Billion Years of Microbial Evolution*. New York: Summit Books, 1986.

Milne, David, David Raup, John Billingham, Karl Niklaus, and Kevin Padian. *The Evolution of Complex and Higher Organisms*, NASA SP-478. U.S. Government Printing Office, 1985.

Mallove, Eugene, Robert L. Forward, Zbigniew Paprotny, and Jurgen Lehmann. "Interstellar Travel and Communication: A Bibliography," Journal of the British Interplanetary Society, Vol. 33, No. 6, June 1980.

Mallove, Eugene, Mary M. Connors, Robert L. Forward, and Zbigniew Paprotny. *A Bibliography on the Search for Extraterrestrial Intelligence.* NASA Reference Publication 1021, March, 1978.

McDonough, Thomas R. *The Search for Extraterrestrial Intelligence: Listening for Life in the Cosmos.* New York: John Wiley & Sons, 1986.

Morris, Richard. *Time's Arrow: Scientific Attitudes Toward Time.* New York: Simon and Schuster, Inc., 1984.

Morrison, Philip, Phylis Morrison, and the Office of Charles and Ray Eames. *Powers of Ten: About the Relative Size of Things in the Universe.* San Francisco: W. H. Freeman and Company, 1982.

Murchie, Guy. *Song of the Sky.* Boston: Houghton Mifflin Company, 1954.

———. *Music of the Spheres.* Boston: Houghton Mifflin Company, 1961.

———. *The Seven Mysteries of Life.* Boston: Houghton Mifflin Company, 1978.

Nagel, Ernest and James R. Newman. *Gödel's Proof.* New York: New York University Press, 1958.

Pagels, Heinz R. *The Cosmic Code: Quantum Physics and the Language of Nature.* New York: Simon and Schuster, Inc., 1982.

———. *Perfect Symmetry: The Search for the Beginning of Time.* New York: Simon and Schuster, Inc., 1982.

Pais, Abraham. *Subtle is the Lord: The Science and the Life of Albert Einstein.* Oxford, Great Britain: Oxford University Press, 1982.

Papagiannis, Michael D. ed. *The Search for Extraterrestrial Life: Recent Developments, International Astronomical Union Symposium No. 112.* Dordrecht, Holland: D. Reidel Publishing Company, 1985.

Ponnamperuma, Cyril and A. G. W. Cameron, eds. *Interstellar Communication: Scientific Perspectives.* Boston: Houghton Mifflin Company, 1974.

Poundstone, William. *The Recursive Universe: Cosmic Complexity and the Limits of Scientific Knowledge*. Chicago: Contemporary Books, 1985.

Premack, David. *Gavagai! Or the Future History of the Animal Language Controversy.* Cambridge, Massachusetts: M.I.T. Press, 1986.

Prigogine, Ilya and Isabelle Stengers. *Order Out of Chaos: Man's New Dialogue with Nature*. New York: Bantam Books, 1984.

Reeves, Hubert. *Atoms of Silence: An Exploration of Cosmic Evolution*. Cambridge, Massachusetts: M.I.T. Press, 1981 (original edition in French).

Rood, Robert T. and James S. Trefil. *Are We Alone?* New York: Charles Scribner's Sons, 1981.

Rowan-Robinson, Michael. *The Cosmological Distance Ladder: Distance and Time in the Universe*. New York: W. H. Freeman and Company, 1985.

Sagan, Carl and I. S. Shklovskii. *Intelligent Life in the Universe*. San Francisco: Holden-Day, Inc., 1966.

Sagan, Carl, ed. *Communication with Extraterrestrial Intelligence: CETI*. Cambridge, Massachusetts: M.I.T. Press, 1973. New York: Doubleday, 1973.

Sagan, Carl. *The Dragons of Eden: Speculations on the Evolution of Human Intelligence*. New York: Random House, 1977.

Sagan, Carl. *Broca's Brain: Reflections on the Romance of Science*. New York: Random House, 1978.

Sagan, Carl, F.D. Drake, Ann Druyan, Timothy Ferris, Jon Lomberg, Linda Salzman Sagan. *Murmers of Earth: The Voyager Interstellar Record*. New York: Random House, 1978.

Sagan, Carl. *Cosmos*. New York: Random House, Inc., 1980.

Schneider, Stephen H. and Randi Londer. *The Coevolution of Climate & Life*. San Francisco: Sierra Club Books, 1984.

Shapiro, Robert. *Origins: A Skeptic's Guide to the Creation of Life on Earth*. New York: Summit Books, 1986.

Smith, Curtis G. *Ancestral Voices: Language and the Evolution of Human Consciousness*. Englewood Cliffs, New Jersey: Prentice-Hall, Inc., 1985.

Sullivan, Walter. *We Are Not Alone: The Search for Life on Other Worlds*. London and New York: McGraw-Hill Book Company, 1964.

Weber, Renée. *Dialogues with Scientists and Sages: The Search for Unity*. New York: Routledge and Kegan Paul, Inc., 1986.

Weinberg, Steven. *The First Three Minutes: A Modern View of the Origin of the Universe*. New York: Basic Books, Inc., 1977.

Williams, Leonard. *The Dancing Chimpanzee: A Study of the Origins of Primitive Music*. London: Allison & Busby, 1980.

Wilson, Edward O. *Sociobiology: The Abridged Edition*. Cambridge, Massachusetts: Harvard University Press, 1980.

Zuckerkandl, Victor. *Man the Musician*. Princeton: Princeton University Press, 1973.

Journal Articles

COSMOLOGY

Barrow, John D. "Anthropic Definitions." *Quarterly Journal of the Royal Astronomical Society* 24:146–153.

———. "The Invisible Universe." *New Scientist* (30 August 1984): 28–30.

Burns, Jack O. "Dark Matter in the Universe." *Sky and Telescope* (November 1984): 396–399.

———. "Very Large Structures in the Universe." *Scientific American* 255, no. 1 (July 1986): 38–47.

Carr, B. J. and M. J. Rees. "The Anthropic Principle and the Structure of the Physical World." *Nature* 278 (12 April 1979): 605–612.

Chincarini, Guido and Herbert J. Rood. "The Cosmic Tapestry." *Sky and Telescope* (May 1980): 364–371.

Collins, C. B. and S. W. Hawking. "Why Is the Universe Isotropic?" *The Astrophysical Journal* 180 (1 March 1973): 317–334.

Crease, Robert and Charles Mann. Review of *The Anthropic Cosmological Principle*. *Science '86* (June 1986): 75–76.

Davies, Paul. "Relics of Creation." *Sky and Telescope* (February 1985): 112–115.

———. "New Physics and the New Big Bang." *Sky and Telescope* (November 1985): 406–410.

———. "The Arrow of Time." *Sky and Telescope* (September 1986): 239–242.

DeLapperant, Valerie, Margaret J. Geller, and John P. Huchra. "A Slice of the Universe." *The Astrophysical Journal* 302, L1-L5 (1 March 1986).

DeWitt, Bryce S. "Quantum Gravity." *Scientific American* (December 1983): 112–129.

Dicke, R. H. and P. A. M. Dirac, Letters to the Editor, "Dirac's Cosmology and Mach's Principle." *Nature* 192 (4 November 1961): 440–441.

Ferris, Timothy. Review of *The Anthropic Cosmological Principle* by J. D. Barrow and F. J. Tipler. *New York Times Review of Books* (16 February 1986): 20–21.

Finkbeiner, Ann. "A Universe in Our Own Image." *Sky and Telescope* (August 1984): 107–111.

Gale, George. "*The Anthropic Principle*." *Scientific American* (December 1981): 154–171.

Gardner, Martin. Review of *The Anthropic Cosmological Principle* by J. D. Barrow and F. J. Tipler. *The New York Review of Books* (8 May 1986): 22–25.

Guth, Alan H. and Paul J. Steinhardt. "The Inflationary Universe." *Scientific American* (May 1984): 116–128.

Freedman, Daniel Z. and Peter van Nieuwenhuizen. "Supergravity and the Unification of the Laws of Physics." *Scientific American* (February 1978): 126–143.

Leslie, John, III. "Anthropic Principle, World Ensemble, Design." *American Philosophical Quarterly* 19, no. 2: 141–151.

Linde, Andrei, "The Universe: Inflation Out of Chaos." *New Scientist* (7 March 1985): 14–18.

Maddox, John. "New Twist for Anthropic Principle." *Nature* 304 (2 February 1984).

Pagels, Heinz R. "A Cozy Cosmology." *The Sciences* (March/April 1985): 34–39.

Press, William H. "A Place for Teleology?" Review of *The Anthropic Cosmological Principle* by J. Barrow and F. J. Tipler. *Nature* 320 (27 March 1986): 315–316.

Smith, David H. "The Inflationary Universe Lives." *Sky and Telescope* (March 1983): 207–210.

Thomsen, Dietrick E. "The Quantum Universe." *Science News* 128 (3 August 1985): 72–74.

Trefil, James. "The Accidental Universe." *Science Digest* (June 1984): 53–55, 100–101.

Tully, R. Brent. "Unscrambling the Local Supercluster." *Sky and Telescope* (June 1982): 550–554.

Weinberg, Steven. "Origins." *Science* 230, no. 4721 (4 October 1985): 15–18.

Wheeler, John Archibald. "Hermann Weyl and the Unity of Knowledge." *American Scientist* (July-August 1986): 366–375.

LIFE

Ball, John A. "Memes as Replicators." *Ethology and Sociobiology* 5 (1984): 154–161.

Further Reading

Cairns-Smith, Graham. "Signs of Life." *New Scientist* (2 January 1986).

———. "The First Organisms." *Scientific American* (June 1985): 90–100.

Coffey, Enrico J. "Life: Its Definition and Some Consequences." *Journal of the British Interplanetary Society* 29, no. 10 (October 1976): 633–640.

Field, Richard J. "Chemical Organization in Time and Space." *American Scientist*, 73, no. 3 (March 1980): 142–149.

Matloff, G. L. and Mallove, E. F. "The First Interstellar Colonization Mission." JBIS, 33, no. 3 (March 1980): 84–88.

Matloff, G. L. and Mallove, E. F. "Solar Sail Starships: The Clipper Ships of the Galaxy." JBIS, 33, no. 9 (September 1981): 371–380.

Matloff, G. L. and Mallove, E. F. "The Interstellar Solar Sail: Optimization and Further Analysis." JBIS, 36, no. 5 (May 1983): 201–209.

Matloff, G. L. "Beyond the Thousand-Year Ark: Further Study of Non-Nuclear Interstellar Flight." JBIS, 36, no. 11 (November 1983): 483–489.

Matloff, G. L. "Interstellar Solar Sailing: Consideration of Real and Projected Sail Materials." JBIS, 37, no. 3 (March 1984): 135–141.

Matloff, G. L. "The State of the Art Solar Sail and the Interstellar Precursor Mission." JBIS, 37, no. 11 (November 1984): 491–494.

Scott, Andrew. "Update on Genesis." *New Scientist*, (2 May 1985): 30–33.

"Second Symposium of Chemical Evolution and the Origin and Evolution of Life." NASA Ames Research Center, Moffett Field, CA, (July 23–26, 1985).

Sagan, Carl and William I. Newman. "The Solipsist Approach to Extraterrestrial Intelligence." *Quarterly Journal of the Royal Astronomical Society*, 24, no. 2: 113–121.

Stebbins, G. Ledyard and Francisco J. Ayala. "The Evolution of Darwinism." *Scientific American*, (July 1985): 72–81.

DESTINY

Asimov, Isaac. "The Last Question" in *Nine Tomorrows*. Fawcett Publications, Greenwich, Connecticut (1959).

Bennett, Charles H. and Rolf Landauer. "The Fundamental Physical Limits of Computation." *Scientific American*, (July 1985): 48–56.

Bohm, David and Renée Weber. "The Physicist and the Mystic—Is a Dialogue Between them Possible?" *ReVision Journal*, (1981).

Darling, David J. "Deep Time: The Fate of the Universe." *Astronomy* 14, no. 1 (January 1986): 6–13.

Dicus, Duane A., John R. Letaw, Doris C. Teplitz, and Vigdor L. Teplitz. "The Future of the Universe." *Scientific American*, 248, no. 3 (March 1983): 90–101.

Dyson, Freeman J. "Time Without End: Physics and Biology in an Open Universe." *Reviews of Modern Physics*, 51, no. 3 (July 1979): 447–460.

Frautschi, Steven. "Entropy in an Expanding Universe." *Science*, 217, no. 4560 (13 August 1982): 593–599.

Margulis, Lynn and Dorion Sagan. "Strange Fruit on the Tree of Life: How Man-Made Objects May Remake Man." *The Sciences*, (May/June 1986): 38–45.

Trefil, James S. "How the Universe Will End." *Smithsonian* no. 3, (June 1983): 73–83.

Vila, Samuel C. "Survival of the Earth and the Future Evolution of the Sun." *Earth, Moon and Planets*, 31 (1984): 313–315.

INDEX

Index

Index

Index

Index